Order and Revolt

Debating the Principles of Eastern and
Western Social Thought

Edited by Wayne Cristaudo,
Heung Wah Wong, and Sun Youzhong

Los Angeles

Order and Revolt: Debating the Principles of Eastern and Western Social Thought
Copyright © 2014
By Wayne Cristaudo, Heung Wah Wong, and Sun Youzhong

Distributed by Transaction Publishers
10 Corporate Place South, Suite 102
Piscataway, NJ 08854

All rights reserved. Exclusive English language rights are licensed to Bridge21 Publications, LLC. No part of this book may be used or reproduced in any matter whatsoever without written permission from the publisher except in the case of brief quotations embodied in critical articles and reviews.

For information contact Bridge21 Publications, LLC, 11111 Santa Monica Blvd, Suite 220, Los Angeles, CA 90025.

Published in the United States
Cover Design by Chi-Wai Li
Copyedited by Peg Goldstein
ISBN 978-1-62643-004-4 Paperback /
     978-1-62643-005-1 Electronic book text

Library of Congress Control Number: 2013948949

Contents

Preface 5
   Wong Heung Wah
Introduction 7
   Wayne Cristaudo

CHAPTER 1
Holistic Thinking and the Reconstruction of the
Notions of Harmony and Spontaneity,
Order and Revolt 19
   Roger T. Ames

CHAPTER 2
What's Wrong with John Dewey and Confucianism and
What's Right with Lao Tzu according to Eugen
Rosenstock-Huessy 35
   Wayne Cristaudo

CHAPTER 3
Mohism, Standards, and Social Order 53
   Donald Sturgeon

CHAPTER 4
Hegel, Lao Tzu, and Bohr: The Merging of Traditions 67
   Robert Elliott Allinson

CHAPTER 5
Justice and Confucian Harmony 85
   Han Rui

CHAPTER 6
Two Concepts of Order:
An Essay on Harmony and Order versus Spontaneity
and Revolt in Western Thought 101
   Hélène Landemore

CHAPTER 7
Japan: The Romanticist Revolt against the Empire 119
  Alexander Dolin

CHAPTER 8
Diderot's Energistic Philosophy and the
Sublime in Evil 141
  Miran Bozovic

CHAPTER 9
Saintly Rebels: Gandhi, the Emir Abdel Kader,
and the Philosophy of Positive Passivity 155
  Waddick Doyle

Endnotes 171
Index 193
About the Contributors 197

Preface

*We Need John Dewey to Understand
Confucius and Vice Versa.
Or Philosophers Need Anthropologists and Vice Versa*

Wong Heung Wah

It is my great honor to be invited to edit this book with Professor Wayne Cristaudo and Professor Sun Youzhong at the later stage of this publication project. As an anthropologist who received almost no serious training in philosophy, I am not sure whether I deserve such a great honor. Anthropologists have devoted themselves to understanding "the other," which I believe necessarily involves cross-cultural comparisons. Dialogue with other cultures is the discipline's essential skill. In other words, the general goal of this book, which is to compare the so-called holistic model of Oriental thought and the Western mode of revolutionary tradition, is anthropologists' normal practice. To understand the other, anthropologists have to take two basic steps. The first step is to involve ourselves deeply in the culture we are to understand through long-term ethnographic fieldwork. The second step is to lie back and understand other cultures *creatively* from an external vantage. That external vantage can be our own culture or another culture. Without these two steps, the task of understanding other cultures cannot really be completed. We need another culture to understand other cultures; and in the context of this book, we need Confucius's harmonious thought to understand John Dewey's pragmatism and vice versa. The same might be said of the much less known relationship between Lao Tzu and Eugen Rosenstock-Huessy, with each one posing a challenge to the more harmonious functionalist view of society provided by Dewey and Confucius. They need each other to achieve mutual understanding. That is to say, comparing Confucius with John Dewey, or Rosenstock-Huessy to Lao Tzu, is not just to point out their differences; through such differences we can better understand

Confucius *and* John Dewey, Lao Tzu *and* Rosenstock-Huessy, the Oriental harmonious whole *and* the Western mode of revolutionary thought. It is in this sense that this philosophy book is also an anthropological exercise, and what the philosophers are doing here is also relevant to the work of their anthropology colleagues. This may be one reason I was invited to join this publication.

Although I am not interested in commenting on the abstract relevance of Confucius in the current sociopolitical situation of China, I think that Confucius's tradition of harmony and order does serve as a mode of thought that structures contingent events in particular ways and thus results in particular historical paths of China, which further makes China historically distinct from its counterparts in the West. It is in this general sense that China's historical processes are guided by Confucius's harmonious whole. Confucius's thoughts are therefore relevant in understanding China. A book like this publication project is extremely useful to anthropologists studying China. Confucius's harmonious whole, however, cannot be deterministic because the events themselves, though contingent or even chaotic, can, as many previous ethnographic studies of China have effectively showed, very often impact and thus transform the general historical path of China. It follows that an anthropology of events is also necessary in understanding the historical courses of China. Perhaps this is another reason I was invited to participate in this book project.

Introduction

Wayne Cristaudo

To the extent that Western political philosophy can be said to begin with Plato's *Republic*, we can rightly say that it originates in the search for political and social harmony. In the East, although Confucius's reflections and instructions for political order lack the speculative dimension that enables Plato to contemplate the best political reality as a pure form or idea from which he may assess the relative disorder of existing political life, Confucius is nevertheless also driven by the question of how to achieve social and political harmony. While not denying the important differences that evolved in China and the West that may be traced back to the peculiar speculative and methodological fork that characterizes Western philosophy, Confucius and Plato and their followers do share a fundamental conviction that the preservation of social order and harmony is the central task of political wisdom or philosophy. But there is another tradition, one that is also common to both East and West. That is a tradition that emphasizes freedom over harmony—or to say it another way, a tradition that sees that the danger of harmonizing potentially conflictual relationships and forces is that it limits the creation of things ex nihilo. Such a tradition involves a certain kind of faith—a faith in the prospect that the unleashing of spontaneous energies will create a better world than one that seeks to harmonize already existing potencies. Revolutions are invariably acts of such faith. And neither China as it exists today nor the West can be understood unless one considers their respective revolutionary underpinnings. Revolutions, though, are the antithesis of harmony, at least of those harmonious orders that obstruct forces striving to shake up an old world and replace it with a new one. Eventually, though, revolutions settle down, new orders with their own dynamics are established, and new ruling elites, in their turn, seek to harmonize the forces under their control. Thus it is that the reality we are part of is one in which such fundamental concepts as harmony and order, spontaneity and revolt are intrinsic to political philosophy.

This volume is an attempt at an East/West exploration of the significance of those concepts within their respective traditions. In this respect the volume is intended as a contribution to a much larger dialogue about values taking place between East and West. I also hope it might be a contribution to dealing with challenges and problems that are common to East and West. Certainly my own motivation, which expressed in a dialogue with coeditor Sun Youzhong helped give birth to this volume, was a problem I see as a global one, and the problem is encapsulated in a question: In spite of the downfall of state-regulated economies, in spite of the ensconcing of rights-based liberalism in much of the industrial world—which of course does not include China—are we living in an age that is becoming increasingly totalitarian? More sharply focused and more conspicuously aligned with the topic of this volume: Is the search for greater social and political harmony contributing to a more totalizing and controlling kind of world that threatens to suffocate spontaneity? Readers of Adorno or even Weber might be sympathetic to this question, for they at least are less likely to consider Soviet or Nazi totalitarianism to be exhaustive. The kind of totalitarianism incorporated in the above question is one in which human beings are increasingly cogs in a vast machine, a machine that may well be dedicated to profit and comfort and that, unlike Nazism or Soviet-styled communism, has no need or desire to monopolize the political process, conquer other lands, or simply exterminate the chosen enemy. Rather the above question is directed at the extent of enmeshment of rules and laws that are intended to produce a benign order—a great harmonious society that conforms to the ideas and plans of a group of people who ostensibly mean well for everybody but that ultimately suffocates the human creativity and expression of vital powers that surface and become active only in climates of freedom.

I do not know the answer to the question—that is, I do not know if the tendencies that to me smack of totalitarianism will be victorious, but I am deeply sympathetic to the idea that the conscious attempt to create ever more order and harmony may be contributing to a way of life where we become ever more machinelike. To an important extent, this fear was one of the most dominant fears of the twentieth century, articulated repeatedly by artists and writers such as the Dadaists, expressionists, surrealists, and situationists and expressed in popular music by such figures as Bob Dylan, Pink Floyd, Devo, and Radiohead. Just as the modernist dream first finds its full-blown modernist formulation in Descartes (and Bacon), the pedigree of skepticism toward the neat symmetries between comfort, mechanization, and freedom that included thinkers as diverse as the Romantics, Tocqueville, Kierkegaard, and Nietzsche can be traced at least as far back as Descartes's nemesis, Pascal. And even if, as seemed tenable until Islamism reminded the world otherwise, the rule of law and free enterprise emerged from the ideological wars not only as natural allies but as seemingly unassailable ingredients of the best possible future for everybody, the fear of the buttresses of modern order forming an asphyxiating totality has never been completely assuaged by economic boom times or ideological victories.

This was, as I have suggested, the opening question that lay behind the dialogue between Sun Youzhong and me, and the conversation that led us to hold a conference at the University of Hong Kong on the topic of this volume. I had just completed a large book on the Jewish philosopher Franz Rosenzweig and a relatively unknown thinker, Eugen Rosenstock-Huessy, who in 1946 wrote a highly provocative chapter that argued that the philosophies of John Dewey and Confucius share some deep affinities, which he saw as totalitarian in the sense alluded to above. He argued that they shared a kind of ultrafunctionalism, not very different in fact from the kind of reasoning that has become ensconced in the managerial revolution that has been sweeping part of the globe for twenty years or so now. (He was, I might add, an opponent of managerialism avant la lettre.) Against them he appealed to what was invaluable and enduring in what he saw as Lao Tzu's refusal to be caught up in the Confucian functionalist social vision. And he integrated Lao Tzu's thinking with his own emphasis on the importance of spontaneity, of things unplanned, of things coming literally out of nothing, and with his reading of the world we now inhabit together as the end result—what he called in an early book "the wedding" of war and revolution.

Sun Youzhong, on the other hand, is an admirer of John Dewey, and when we first spoke of these matters, he told me that he had recently attended a conference on Confucius and Dewey. He believed in the seriousness of the issues we were discussing, and while he was not completely dismissive of Rosenstock-Huessy's arguments, he thought that Confucius and Dewey have much to offer in creating a better world. He was also sure that Confucius and Dewey scholars would welcome such a dialogue around the concepts of spontaneity and revolution, harmony and order, thus also giving them an opportunity to present their case in light of these kinds of criticism.

Ultimately what interested me and Sun Youzhong was not so much whether Rosenstock-Huessy was or was not an adequate interpreter of Dewey and Confucius (we were not primarily interested in scholarship that simply demonstrates that Y does not understand X) but the issues that Dewey, Confucius, and Rosenstock-Huessy were addressing and how they addressed them. For my part I welcomed a forthright exchange of points of view. And the exchange at the conference was wonderfully forthright. For example, Roger Ames, the celebrated North American Confucian scholar who has a paper in this collection and who wrote a very important book with David Hall, *Democracy of the Dead: Dewey, Confucius, and the Hope for Democracy in China*,[1] arguing for the value of fusing Dewey and Confucius for China's future, said he thought Rosenstock-Huessy was wrong on Dewey, Confucius, and Christianity (the tradition Rosenstock-Huessy appeals to against Dewey and Confucius)—though I thought everything in Ames's paper, and indeed in all papers discussing Confucianism and/or Dewey at the conference, confirmed precisely what Rosenstock-Huessy was saying! Ultimately, though, I think the question of whether Rosenstock-Huessy's or Roger Ames's assessment of Dewey and Confucius is more accurate is not nearly as important as the questions and dialogue that may emerge from closer scrutiny of the concepts of order and harmony and spontaneity and revolt that form the basis of his provocation. This was really the point of the dialogue.

Anyone familiar with twentieth-century Western philosophy is aware of the antitotalizing motif shared by thinkers such as Benjamin, Rosenzweig, Levinas, and Derrida, but as Robert Allinson indicates in his "Hegel, Lao Tzu, and Bohr: The Merging of Traditions," there remains a hankering for holism and dialectical reconciliation shared by many Westerners (such as Niels Bohr), who find in both Confucius and Lao Tzu a source of inspiration for an integrated view of life in which opposites each have their part to play. On the other hand, Han Rui argues in "Justice and Confucian Harmony," in a paper that draws heavily upon John Rawls, that the complexities of modern life create rifts that simply may not be harmonized but that nevertheless may be dealt with justly. In this respect she identifies serious limits to what Confucianism has to offer the modern world. Interestingly, Han Rui too appeals to Dewey, but unlike Ames, she does so to highlight the difference between his kind of liberalism and Confucianism.

When Professor Sun and I set upon this undertaking, we also decided that the examination of concepts should be extended beyond Confucius and Dewey to include contributors who could throw more cross-cultural light on the concepts we thought should form the basis of the discussion. Thus Donald Sturgeon, a PhD candidate at the University of Hong Kong, contributed a chapter on Mohism, a philosophy that had little traditional impact upon China but that was conceived as an alternative to Confucianism. Moreover, with the contributions of Alexander Dolin's "Japan: The Romanticist Revolt against the Empire" and Waddick Doyle's "Saintly Rebels: Gandhi, the Emir Abdel Kader, and the Philosophy of Positive Passivity," we had two excellent papers that may broadly be grouped under the East perspective and that are very pertinent to the discussion. From the Western side of things, we also were fortunate to receive Hélène Landemore's "Two Concepts of Order: An Essay on Harmony and Order versus Spontaneity and Revolt in Western Thought," which is an excellent analysis of the tensions surrounding the central concepts in Western philosophy in this volume. I think she has presented a powerful and persuasive analysis clarifying the dynamism at the heart of the Western tradition. Also extremely relevant for our purposes is Miran Bozovic's "Diderot's Energistic Philosophy and the Sublime in Evil." He goes even further than Landemore in that his analysis brings up the modern Western fascination with the demonic and how the dynamism of modernity needs to be understood as the harnessing of the energy of evil. A commonplace erroneous claim is that European culture rests upon its privileging of rational order, reason writ large. Jacques Derrida, whose position is much more nuanced than that frequently ascribed to him, is often cited for his attack upon European/Western logocentrism. Funnily enough, the one current of the West that would most *seem* to fit this picture, the Enlightenment, does not completely fit into the template of such a critique. Denis Diderot, as Bozovic's introduction to his essay correctly points out, is in a line of thinkers for whom energy rather than rational order is predominant—yet he is also a key figure in the Enlightenment. In 1797 the German critic Friedrich Schlegel noted the tendency of modern poetry to Satanism [2] —a tendency that would become fully blown with Charles Baudelaire's *Flowers of Evil*. Bozovic's paper focuses on Diderot rather than on that tradition as a whole, but the reasons provided by Diderot for evil's sublimity are far from idiosyncratic and provide an important cipher about what may be called the power of the negative of harmonious order. The close alliance between Enlightenment and revolution finds its point of contiguity precisely in Diderot's fathoming of the sublimity of evil and the primacy of energy over moral order.

Although, as I have suggested, we were not limiting our discussion to Chinese perspectives, I think it fair to say that what makes the discussion particularly important now is the acceleration of China–West relations taking place over the last decade or so. And if I may, I would like to use this introduction to expand on the rationale behind my own thinking, which also makes the case for the timeliness of this topic.

How we came to have the world we have is a question with a very complicated answer, but I venture to say that had it not been for the string of revolutions in the West, culminating in the Russian Revolution, it is highly unlikely that we would have a world even remotely like this one. There is nothing triumphalist in this statement—revolutions are not only terrifying, but they occur because of what is wrong in a tradition, not what is right in it. However, revolutions invariably create hitherto unforeseen ways of *making* men and women. Revolutions inevitably smash inherited social foundations and traditional sources of appeal while establishing new pathways into the future. While those new pathways are without precedent, inevitably—as Montesquieu grasped from his observations of the English Revolution and as he explained while trying to prevent a French Revolution—the heirs of the revolution find they must restore some, and even a large part, of what the revolution hoped to consign to oblivion. Yet what is restored functions in a thoroughly new context, and thus the potencies it discharges are of a very different order and take on a different array of effects than what they had originally unleashed or been designed for. One example of this process is Confucianism. The Confucius who has been restored from the Communist Party's earlier banishment finds himself required to be a teacher to a people struggling to harmonize a market-driven economy aiming at high levels of personal wealth for all its members. That is, Confucius is being called upon to help stop the social disintegration that invariably accompanies every nation drastically and rapidly changing the nature of its social relationships. I am rather skeptical of what this restored Confucius can do, but others in this volume find there is still much to draw from his well for the future of China, and possibly even beyond. They may well be correct. In any case, restorations happen when revolutions run out of steam, when the only way forward is back. Heritages are forces that operate behind our backs—revolutions occur when heritages break backs, when they no longer provide key members of a social group (usually young, disillusioned, educated, and daring men) with a future they think is worth having. What men like Sun Yat Sen, Chiang Kai Chek, and Mao Tse Tung had in common was that in trying to save China from a heritage that they saw as no longer offering a desirable future, they all took their cues from revolutionary ideas, processes, and institutions that came from nonautochthonous heritages.

To be sure, the specific set of revolutionary ideas they took their cues from were not the same, though broadly speaking they came from the same "other"—the West. And just as the revolutions they took their cues from occurred in different locations and times, the revolutionaries attached themselves to the powers of different times in order to deal with their own times. Moreover, the fact that China today is a communist nation-state with a market-based economy points to the fact that it has entered into the same revolutionary dynamism that first took hold of the West and now has spread globally—for every one of the terms I have just used to describe China had a revolutionary origin. One could readily add that the political leadership in China is based upon a model that was imported from the Russian Revolution, but its economic base is a product of the English Revolution, its international political shape is from Westphalia (an indirect result of the Reformation, or German Revolution), and it is presently having to deal with demands of representation and internal nationalist dissent that are deeply reminiscent of the French Revolution.

In making this point about the Western origin of important dimensions of China's present social basis, I am not precluding the fact that China will and already is undoubtedly contributing to a planetary future; nor am I denying for a moment that China infuses what it "borrows" with its own cultural characteristics and that what comes out of this hybrid is something (at least in some ways) unique. But what I want to emphasize is simply that revolution is intrinsic to the reality of China every bit as much as it is to that of every Western country and that the revolutions within the West are intrinsic to how China must find its own internal harmony as well as to how it must relate to the international order.

It was Rosenstock-Huessy's *Out of Revolution* (which was brought to my attention by another fascinating book, *Law and Revolution*, by his student the legal scholar Harold Berman) in which I first saw the case made that the great revolutions of the West formed a sequence in which the achievements of earlier revolutions spread throughout Europe and thus led to a transformation of all European nations, even if as a reaction, such as with the counter-Reformation. Thus, while local conditions play a key part in where a particular revolution breaks out, the revolution is itself invariably against circumstances and conditions that are not merely local. This is why, for example, the Russian Revolution was welcomed by progressives throughout Europe and, even more importantly, outside of Europe and why the French Revolution was greeted so enthusiastically by men like Kant and Blake. Baldly put, the revolutions of Europe have made it meaningful to talk of human progress; this does not mean that progress is either automatic, straightforwardly linear, or immune from catastrophe, but it does mean that certain events have been so traumatic that social collectives have learned certain lessons that have been incorporated into institutions so that these lessons can be passed down to later generations. Sometimes battles will have to be refought, and victory is never assured, but certain things have changed irrevocably as certain social roles and types have either disappeared completely—the feudal lord and serf—or have been so radically transformed that they bear only a vague resemblance to their distant ancestors—European monarchs and titled nobles, for example. To say it again, revolutions establish new beginnings and exterminate parts of a social order. Indeed, they often condemn the very bulwarks and buttresses of a social order. They also transform the very nature of the relationship between the living and the dead. A postrevolutionary society—which is to say all societies on the planet today, which for better or worse are heirs to the world wars (both of which would have never come about were it not for the commercial and scientific revolutions, nation-states, and the nationalism that Napoleon extracted from the French Revolution and spread throughout Europe)—is necessarily one that no longer venerates entire *types* of ancestors. (Of course, a certain kind of moral character may still be "sanctified" beyond his or her type by later generations.) Another way of saying this is that social harmony may mean a benign present, which may also be the enemy of progress and human possibility, while revolution may be the horror of the present that is the gift to the future.

Consider Aristotle, for example. His view of politics is thoroughly decent. He looks at how society as a whole may best prosper and how each member of a polity may benefit in accordance with his or her social roles. He grasped the terrible dangers that face city-states where there is not a strong enough middle class to ameliorate the tensions between the wealthy and the poor. Unlike Plato, who simply established an ideal from which to observe the pathologies of a regime, Aristotle proceeded like a doctor when he came to investigate the various Greek constitutions—he would always look at how a diseased social order could become more healthy, realizing that for many polities there was no question of real happiness and human flourishing for most members. More generally, he also considered the roles of slaves and women, and his discussion is humane—for he never urges anything but humane treatment of the slave by nature (on slavery he distinguishes between those who are slaves due to misfortune and those who are slaves by nature) and women, whom unlike slaves by nature do not lack a deliberative faculty, but women do lack the authority to proceed without male direction. His decency is a response to his observations, but his observations are also a confirmation of a restrictive order, which if maintained would condemn women and slaves by nature to roles of subordination into perpetuity. The question of living in service with others is a complicated one, and I think the Reformation conception of freedom as bondage is far truer to the more common modern concept of non-impingement upon one's will—what Isaiah Berlin called negative liberty. But one great achievement of modernity is the fluidity of its social roles. To be sure, there are many ways within "free societies" for lives to be wrecked, but that is not an argument for the restoration of slavery or the oppression of women. Aristotle's social analysis, then, for all its humaneness, is the opposite of the progressive position. And it is his emphasis upon social harmony that makes him an enemy of progress. We who have the advantage of thinking about society on the basis of revolutionary changes can readily see that the harmony he preserves deserved to perish. (It is a fact that such a view of revolutionary progress, as opposed to the rebellious overthrow of the wealthy by the poor, lay outside the Greek experience.) Today to say that slavery should be prohibited and that women should be citizens is as platitudinous as it is contrary to what Aristotle thought was possible. And of course, the dissolution of slavery and the expansion of the franchise to include women were achievable within certain institutional parameters, but those parameters were expanded precisely by those who attacked slavery and women's servitude. Ultimately, in spite of Aristotle's arguments that preserve slavery and women's dependency, it would be truer to say that Aristotle wished to preserve harmony and sought to instruct citizens to live decently and fairly rather than that he wanted to suppress women or slaves. But the order he thought about and wished to preserve had its own in-built suppressive mechanisms, and Aristotle's thought does not contribute to attacking those mechanisms. In sum, Aristotle's notion of harmony can be seen as seriously flawed.

What the above also suggests is that we may think of harmony in a number of ways, but two particularly interest me here. One is the harmony of the type I have been alluding to in my discussion of Aristotle. This is the harmony of what is, the harmony that does the best it can with the materials at hand. It is a harmony that preserves a specific sacrificial order. (All orders, I suggest, are built upon sacrifice, and in antiquity, though not in modernity, all social roles require sacrificial conduct from members.) The problem with that order, as I have suggested already, is that the sacrifices that may be necessary to preserve that *particular order* may not be that worthwhile, because that particular order may be one that thwarts all manner of human potential. This may not mean much to those who live well within the order, but it does to those whose sacrifice is early death, the possibility of famine, cruel punishment, and cruel treatment from superior classes. Then there is another kind of harmony: the harmony of the future. In the Western tradition, this is closely bound with the idea of the messianic. This too is a particular understanding of harmony, but it is a radical harmony—one that bursts relationships that seem necessary and intrinsic and replaces them with possibilities heretofore unimagined. This is a harmony so contrary to expectation that the lion will lie down with the lamb. The difference between this kind of harmony and the kind that requires reconciling the forces existing at any one time in a social body is that the future harmony is premised upon judgment day. That is to say, it is premised upon the destruction of types and practices. It is true to say that the revolutionaries of the West have always looked to that kind of harmony, even if they have all suffered under the conceit that their revolution was the event that would introduce it. It is also true to say that in every process of revolution, there is a moment where the revolution seems totally defeated—all seems to have been a meaningless squandering of life. The collapse of the Soviet Union would suggest that the Russian Revolution amounted to nothing—but did it? No doubt the use of the revolution to maintain Russia's empire fell, as did the idea of communism as total state control of the means of production, but the Russian Revolution provided, for better or worse, the political model that provided a *faith* for people who had no choice but to develop economically (and thus transform themselves socially), swiftly, and irrevocably. The Russian Revolution certainly never achieved communism, but it just as certainly dragged the Russian Empire into the modern world of industry and atomic power. Marx was absolutely correct when he said that "capital comes dripping from head to toe, from every pore, with blood and dirt,"[3] but it was no less true for socialism. It is all too understandable that the horrific nature of revolutions leads to their moral condemnation. But the fact is that the radical large-scale transformation of social and political institutions—as happens when entire classes and ruling orders must be overthrown to enable the expansion of new or alien forms of social life—is always paid for by blood. (This is not a normative statement.) And in spite of Marx's ostensible anti-idealist anti-moralizing politics, neither liberal nor socialist critics of either the bourgeois-liberal order or the socialist order are able to escape the extramoral logic of social transformations: that all social orders are predicated on selection and sacrifice (these qualities and these peoples are favored over those qualities and those people) and that a great upheaval means rearranging who and what is selected and sacrificed. This is why undoing a revolution is also a great act of chaos. Though with the Russian Revolution, the dismantling of the revolution was far less catastrophic than the original revolution. The society was no longer launching into an almost totally different set of social relations as modernity demands from traditional or agrarian societies, yet the change to the economy was chaotic enough.

In this respect, China's decision not to dismantle its political inheritance from the Russian Revolution, not to create a political vacuum, looks now to have been by far wiser than the Soviet Union's attempt to "liberalize" itself by dismantling itself (China, though, already had set itself on a non-Soviet-styled economic path); of course, after a relatively short period of social and economic chaos that achieved little real emancipation, Putin has succeeded in reestablishing an authoritarian form of politics, which has nothing in common with what the West and Russia's relatively small number of liberals wished for but which is far more economically efficient than the Soviet model. Just as Putin has been unable to completely shake off the heritage of the Soviets, the Soviets themselves, in spite of their antibourgeois rhetoric, found themselves anchored to such anticommunist elements, or bourgeois residues, as the retention of money, marriage (attempts to jettison marriage were dropped very quickly), and bosses. (Instead of being eliminated, they indeed became a "new class," described by Yugoslav dissident Milovan Ðilas in *The New Class*.) Likewise, the factionalism that in the French Revolution had led to parties becoming the basis of revolutionary politics was maintained. And what had been a particular phase of violence in the French Revolution based upon appeals of survival, necessity, and virtue became the standard modus operandi of Soviet states. And when Stalin announced "socialism in one country" he conceded what had been obvious from the socialist divisions within Europe on the eve of the Great War—that he could not escape what was possibly the most decisive revolutionary contribution of the bourgeoisie: the idea of the nation.

Liberal order, as Friedrich Hayek astutely saw, was based on an insight about a new type of order—not what he said was the more traditional order of *taxis*, which is appropriate to a military plan, a top-down organizational model, but a self-ordering order, an unintended design or *catallactic* order. Twentieth-century liberal societies have struggled over the balance between taxis and catallaxy, and it is fair to say that both forms of order have brought certain benefits. But liberal societies have increasingly proven to be "statist" societies. China today is likewise struggling with this balance.

What the West and China have in common is that they are both struggling. This is why both are eager to learn from one another. And part of that learning is about different ways of understanding order and harmony. I have suggested that revolution has been an intrinsic part of the common condition of East and West; others in this volume are interested in holistic views of harmony that may be beneficial not only to China but to the world as a whole. I hope that by considering the meaning of order and harmony, spontaneity and revolt in light of East/West perspectives, this volume might make a small contribution to what each may learn from the other.

Chapter 1

Holistic Thinking and the Reconstruction of the Notions of Harmony and Spontaneity, Order and Revolt

Roger T. Ames

The starting point for the many resonances to be found between the pragmatism of John Dewey and the philosophy proffered in the canons of classical Confucianism is their shared commitment to holistic thinking—what A. N. Whitehead would describe quite simply as a commitment to an aesthetic as opposed to an abstracted rational sense of order.[1] The cosmological assumptions of both Deweyan pragmatism and Confucianism with respect to harmony and spontaneity, order and revolt are grounded in an empirical naturalism that precludes appeal to any independent supernatural or metaphysical sources lying outside of experience and that rules out the possibility of a mind-independent understanding of experience.

In this essay I will first examine the Deweyan notion of experience.[2] It follows from Dewey's understanding of experience as being resistant to any and all forms of dualism—including that of subject and object—that any exploration of Dewey on experience will necessarily require an account of its resolutely subjective dimension. We will then turn to classical Confucian philosophy, and by examining the traditional focus-field understanding of persons and their environments, we will establish a shared aestheticism as an interpretive context for reconsidering and reconstructing an understanding of the relationship between harmony and spontaneity, and between order and revolt.

DEWEYAN PRAGMATISM AND THE HOLISTIC NATURE OF EXPERIENCE

Dewey, in "The Postulate of Immediate Empiricism," is explicit in his insistence that we rely upon our immediate experience as the appropriate resource for philosophical speculation and that we owe it our ultimate allegiance in evaluating these same conjectures. In his own language: "Immediate empiricism postulates that things—anything, everything, in the ordinary or non-technical use of the term 'thing'—are what they are experienced as. Hence, if one wishes to describe anything truly, his task is to tell what it is experienced as."[3]

For Dewey, all experiences, regardless of their ultimate veracity, are equally real. In this claim, Dewey seems to be reiterating William James's understanding of "radical empiricism" that his mentor defined in the following terms: "To be radical, an empiricism must neither admit into its constructions any element that is not directly experienced, nor exclude from them any element that is directly experienced."[4]

Although Dewey is certainly influenced by James, the important difference between James and Dewey is that Dewey wants to reject any notion of primal stuff out of which experience evolves and to assert an ontological parity among all experiences as such:

> There is, then, from the empiricist's point of view, no need to search for some aboriginal *that* to which all successive experiences are attached, and which is somehow thereby undergoing continuous change. Experience is always of *thats*; and the most comprehensive and inclusive experience of the universe that the philosopher himself can obtain is the experience of a characteristic *that*.[5]

Dewey wants to argue for the reality of "that experience" and "experienced as." This is what he means when he claims that "by these words [*that* and *as*] I want to indicate the absolute, final, irreducible and inexpungable concrete quale which everything experienced not so much *has* as *is*."[6]

The contemporary pragmatist Hilary Putnam offers an interpretation of experience that is consistent with Dewey in highlighting the irreducibly subjective dimension of what Dewey calls "that" or "as." In Putnam's rejection of the objective mind-independent world of naive realism, he argues that "elements of what we call 'language' or 'mind' *penetrate so deeply into what we call 'reality' that the very project of representing ourselves as being 'mappers' of something 'language-independent' is fatally compromised from the start.* Like Relativism, but in a different way, Realism is an impossible attempt to view the world from Nowhere"[7]

Putnam continues by insisting that this kind of human penetration and transformation of our environments extends to our attention and valorization of the world in which we live and requires us to accept our own reflexivity as "beings who cannot have a view of the world that does not reflect our interests and values."[8] For Putnam, like Dewey, we are part and parcel of what experience *really* is: "The heart of pragmatism, it seems to me—of James' and Dewey's pragmatism if not of Peirce's—was the supremacy of the agent point of view. If we find that we must take a certain point of view, use a certain 'conceptual system,' when engaged in a practical activity, in the widest sense of practical activity, then we must not simultaneously advance the claim that it is not really 'the way things are in themselves.'"[9]

Dewey, in invoking truisms such as "experience is experience" and "it is what it is," provides us with a philosophical methodology for clarifying the many distinctions that we rely upon to think philosophically: "If you want to find out what subjective, objective, physical, mental, cosmic, psychic, cause, substance, purpose, activity, evil, being, quality—any philosophic term, in short—means, go to experience and see what the thing is experienced *as*."[10]

Of course, in this particular essay, I want to add *harmony, spontaneity, order,* and *revolt* to Dewey's list of philosophical terms, anticipating that they too must be reconstructed when examined through the lens of the wholeness of experience. But to be true to Dewey's own premises, such holism requires that these abstract cosmological notions be understood reflexively and thus be given the concrete context of subjective experience. To provide such a context, I will turn to one of Dewey's most revolutionary theses arising from his immediate empiricism—that is, his relational and transactional understanding of organism and environment when it is applied to the process of becoming a person. Dewey's holism requires that the notion of person be understood as being radically embedded in experience as a vital configuration of aggregating conduct—what Dewey describes as a habitude.

DEWEY AND THE TRANSACTIONAL, GERUNDIVE NOTION OF INDIVIDUALITY

Dewey, in his phenomenology of human conduct, combines the process psychology of William James and the social psychology of George Herbert Mead to locate persons as foci of activity within their natural and social relations. He follows Mead, who insists that "self" is coterminous with the world:

> The self cannot arise in experience except as there are others there. The child experiences sounds, etc., before it has experience of its own body; there is nothing in the child that arises as his own experience and then is referred to the outside things.... Only a superficial philosophy demands the old view that we start with ourselves.... There is no self before there is a world, and no world before the self. The process of the formation of the self is social.[11]

Mead and Dewey too are revolutionary within the Western narrative in dispensing with the "old psychology" that begins from assumptions about a superordinate and discrete *psyche*. In his *Human Nature and Conduct* (perhaps more appropriately titled *Human Nature as Conduct*), Dewey calls into question the separation of "human nature" and "conduct" by disputing the very distinction between them. Indeed, Dewey shaves with Ockham's razor by dispensing with the notion of a superordinate and hence reduplicative person in any of its permutations—soul, self, psyche, mind, character, agency, and so on. He insists that cultivated moral dispositions as habits of conduct are themselves the evolving reference of what it means to become a consummate person. In seeking to overcome the default assumptions that have become entrenched within his own tradition, Dewey is obliquely critical of his mentor, James, who is not always consistent on the status of the individual person:[12] "The doctrine of a single, simple and indissoluble soul was the cause and the effect of failure to recognize that concrete habits are the means of knowledge and thought. Many who think themselves scientifically emancipated and who freely advertise the soul for a superstition, perpetuate a false notion of what knows, that is, of a separate knower."[13]

Dewey instead, in a different language but analogous in many ways to the Confucian notion of a relationally constituted person, which I will describe below, arrives at an understanding of the human being as a dynamic combination of habit and impulse: "Now it is dogmatically stated that no such conceptions of the seat, agent or vehicle will go psychologically at the present time. Concrete habits do all the perceiving, recognizing, imagining, recalling, judging, conceiving and reasoning that is done. 'Consciousness,' whether as a stream or as special sensations and images, expresses the functions of habits, phenomena of their formation, operation, their interruption and reorganization.... A certain delicate combination of habit and impulse is requisite for observation, memory and judgment."[14]

Dewey further insists the initial biological and social conditions situating us within community need to be taught through a robust process of nurturance and growth: "We are born organic beings associated with others, but we are not born members of a community."[15] For Dewey, "individuality" is not quantitative: it is neither a presocial potential nor a kind of isolating discreteness. Rather, it is a qualitative achievement, arising through distinctive service to one's community. Individuality is "the realization of what we specifically are as distinct from others,"[16] a realization that can take place only within the context of a flourishing communal life. "Individuality cannot be opposed to association," says Dewey. "It is through association that man has acquired his individuality, and it is through association that he exercises it."[17] An individual so construed is not a discrete "thing" but a "patterned event," describable in the language of uniqueness, integrity, social activity, relationality, and qualitative achievement.

Dewey appeals to the actual situated human experience as we live it in rejecting the notion of an autonomous "self" that is ostensibly prior to an organic configuration of relations and in questioning the putative priority of self-based instincts to shared cultural life-forms. He is, in his own tradition, a flat-out revolutionary in insisting upon the primacy of the relationally constituted, concrete situation as the garden of social intelligence and the soil that nourishes the blossoming of the moral life. The optimally appropriate response to our ever-present uncertainties, and any confident resolution of these uncertainties, can be negotiated only within the actual circumstances themselves, with any notion of individual agency itself being an abstraction from these circumstances:

> The primary significance of the unique and morally ultimate character of the concrete situation is to transfer the weight and burden of morality to intelligence. It does not destroy responsibility; it only locates it. A

moral situation is one in which judgment and choice are required antecedently to overt action. The practical meaning of the situation ... the action needed to satisfy it—is not self-evident. It has to be searched for.[18]

Dewey, in this social construction of the person, rejects the idea that persons are not only incomplete outside of the association they have with other people but in fact have no point of reference beyond this context. Importantly, in this notion of relationally constituted and emergent "individuality," to say that persons are irreducibly social is not to deny the integrity, uniqueness, and diversity of human beings; on the contrary, it is to affirm the achievement of these conditions. Said succinctly, we are who we are not exclusive of our relations but because of them.

CONFUCIAN COSMOLOGY AND THE FOCUS-FIELD NATURE OF EXPERIENCE

In what way does a person *become* consummately human? This is the perennial Confucian question asked explicitly in each of the canonical *Four Books*. And the answer from the time of Confucius onward continues to be a moral, aesthetic, and ultimately religious one. One *becomes* human by cultivating those thick, intrinsic relations that constitute one's initial conditions and that locate the trajectory of one's life force within family, community, and cosmos.[19] "Cultivate your person"—*xiushen* 修身—the signature exhortation of the Confucian canons—is the ground of the Confucian project of becoming consummate as a person (*ren* 仁): It is to cultivate one's conduct assiduously as it is expressed through those family, community, and cosmic roles and relations that one lives. In this Confucian tradition, we need each other. If there is only one person, there are no persons.[20] Becoming a person is something that we *do* and that we either do together or not at all.

The Great Learning 大學, the seminal, foundational document that sets and anchors this Confucian project early in the tradition, describes the process of becoming human, insisting that it is only through committing oneself to a resolute regimen of personal cultivation that one can achieve the comprehensive intellectual and moral understanding that will make the most of the human experience for each of us in our families and communities. The central message of this terse yet comprehensive document is that while personal, familial, social, political, and indeed cosmic cultivation is ultimately coterminous and mutually entailing, it must always begin from a commitment to personal cultivation. In the enduring language of the text itself:

> The way of achieving greatness through learning lies in demonstrating real personal excellence, in cherishing the common people, and in dedicating oneself to doing what is best. Such a course of learning can only be set once one has made this commitment. Only in having set such a course is one able to find equilibrium, only in having found equilibrium is one able to become self-assured, only in having become self-assured is one able to be deliberate in what one does, and only in being deliberate in what one does is one able to get what one is after. There is the important and the incidental in things and a beginning and an end in what we do. It is in realizing what should have priority that one approaches the proper way (*dao*).[21]

Having endorsed the priority of making a commitment to personal cultivation, the text continues by rehearsing the cosmic proportions of the achievement of the ancient sage-kings once they dedicated themselves to this project:

> The ancients who sought to demonstrate real excellence to the whole world first brought proper order to their states; in seeking to bring proper order to their states, they first set their families right; in seeking to set their families right, they first cultivated their own persons; in seeking to cultivate their persons, they first knew what is proper in their own hearts-and-minds; in seeking to know what is proper in their hearts-and-minds, they first became sincere in their purposes; in seeking to become sincere in their purposes, they first became comprehensive in their wisdom. And the highest wisdom lies in seeing how things fit together most productively.
>
> Once they saw how things fit together most productively, their wisdom reached its heights; once their wisdom reached its heights, their thoughts were sincere; once their thoughts were sincere, their hearts-and-mind

knew what is proper; once their hearts-and-mind knew what is proper, their persons were cultivated; once their persons were cultivated, their families were set right; once their families were set right, their state was properly ordered; and once their states were properly ordered, there was peace in the world.[22]

Each person stands as a unique perspective on and a unique opportunity for family, community, polity, and cosmos, and through a dedication to deliberate growth and articulation, all have the possibility of bringing the resolution of the web of relationships that locate and constitute them within family and community into clearer and more meaningful focus. The "learning" (xue 學) of *The Great Learning* is nothing more or less than the cultivation of productive, transpersonal habits of conduct.[23]

As *The Great Learning* enjoins us, in the singularly important project of becoming consummate persons, we must get our priorities right: "From the emperor down to the common folk, everything is rooted in personal cultivation. There can be no healthy canopy when the roots are not properly set, and it would never do for priorities to be reversed between what should be invested with importance and what should be treated more lightly."[24]

The *Record of Rites* (*Liji* 禮記) version of *The Great Learning* brings this text to its conclusion by declaring that giving priority to achieving personal excellence is wisdom at its best. In its own words: "This commitment to personal cultivation is called both the root and the height of wisdom."[25]

Just as in the life of any plant form, the "root" of personal cultivation and its ultimate product, wisdom, far from being separate and distinct, are perceived as an organic whole that must grow together or not at all.

In appealing to an understanding of Chinese natural cosmology as the relevant interpretive context for this Confucian project, I want to provide a language that will distinguish this worldview from the reductive, single-ordered "one-behind-the-many" ontological model that grounds classical Greek metaphysical thinking, wherein one comes to "understand" the many by knowing retrospectively the foundational and causal ideal that lies behind them. Instead, we find that in Chinese cosmology there is a symbiotic and holistic focus-field model of order that is illustrated rather concisely in the organic, ecological sensibilities of *The Great Learning* 大學. As we have seen, this text begins from the ecological interconnectedness and interdependence of all of the many dimensions of the human experience within which this collaborative process of personal consummation is to be pursued. The Confucian project begins from a recognition of the wholeness of experience and the constitutive nature of relationality entailed by it. Moreover, because each person and event is constituted by an interdependent web of relations, what affects one thing affects all things. Meaningful relations within this family make the entire cosmos more meaningful; relations that remain barren detract from it.

The meaning of the family is implicated in and dependent upon the productive cultivation of each of its members. And by extension, the meaning of the entire cosmos is implicated in and dependent upon the productive cultivation of each person within the family and community, just as an entire symphony lies implicated within each note that expresses it. Personal worth is the source of human culture, and human culture in turn is the aggregating resource that provides a context for each person's cultivation.

Said another way, within the correlative cosmology that serves as context for the development and evolution of Confucianism, nothing happens on its own. Nothing happens unilaterally and in isolation. Physically, breathing is a symbiotic collaboration between lungs and air, seeing between eyes and sun, running between legs and ground, and socially, friendship is a concrete relationship between friend and friend. All activity occurs within a context and is thus necessarily collateral in nature. Unsurprisingly, we find that *all* of the terms of art defining the Chinese natural cosmology are binomial rather than singular, reflecting this ubiquitous collaterality. The vocabulary is transactional and collaborative: "the natural, cultural, and numinous context and humanity" (*tianren* 天人), "the heavens and the earth" (*tiandi* 天地), "forming and functioning" (*tiyong* 體用), "flux and continuity" (*biantong* 變通), "the furthest reaches and beyond" (*taiji/wuji* 太極無極), the *yin* and the *yang* 陰陽, "this particular vital focus and its field" (*daode* 道德), "configuring and vital energy" (*liqi* 理氣), "determinacy and indeterminacy" (*youwu* 有無), and so on. No term can stand alone as an independent, determinative principle. There can be no superordinate and independent "one" in this ecological cosmology—no single cause; no grounding, foundational standard; no one privileged order. Consonant with this observation, the distinguished French scholar Marcel Granet, like many other preeminent Chinese and Western sinologists in their own terms, tells us that "Chinese wisdom has no need for the idea of God,"[26]

While certainly having important theoretical implications, the enduring power of this Confucian project is that it proceeds from a relatively straightforward account of the actual human experience. It is a pragmatic naturalism in the sense that, rather than relying upon metaphysical presuppositions or supernatural speculations, it focuses instead on the possibilities for enhancing personal worth available to us here and now through enchanting the ordinary affairs of the day. A grandmother's love for her grandchild is at once the most ordinary of things and the most extraordinary of things, the most common of things and the most sublime.

Confucius, by developing his insights around the most basic and enduring aspects of the ordinary human experience—family reverence, deference to others, friendship, a cultivated sense of shame, education, community, and so on—has guaranteed their continuing relevance. One characteristic of Confucianism that is certainly there in the words of Confucius himself and that has made his teachings so resilient in the Chinese tradition is its porousness and adaptability. His contribution was simply to take ownership of the cultural legacy of his time, to adapt the wisdom of the past to his own present historical moment, and then to recommend to future generations that they do the same.[27] What in English is called Confucianism, associating this tradition with one person, is in Chinese *ruxue* 儒學—the compounding learning of eighty generations of literati who have participated in the corporate Confucian project.

The personal model of Confucius that is remembered in *The Analects* does not purport to lay out some generic formula by which everyone should live their lives. Rather, the text recalls the narrative of one special person: how he in his relations with others cultivated his humanity and how he lived a fulfilling life, much to the admiration of those around him. We might take liberties and play with the title of *The Analects*, reading "discoursing" (*lunyu* 論語) more specifically as "role-based discoursing" (*lunyu* 倫語). Indeed, in reading *The Analects*, we encounter the relationally constituted Confucius making his way through life by living his many roles as best he can: as a caring family member, as a strict teacher and mentor, as a scrupulous and incorruptible scholar-official, as a concerned neighbor and member of the community, as an always critical political consultant, as the grateful progeny of his progenitors, as an enthusiastic heir to a specific cultural legacy, indeed as a member of a chorus of joyful boys and men singing their way home after a happy day on the River Yi. He offers us historical models rather than principles and exhortations rather than imperatives. The power and lasting value of his insights lie in the fact that, as I will endeavor to show, these ideas are intuitively persuasive and readily adaptable to the conditions of ensuing generations, including our own.

Indeed, invoking the Chinese natural cosmology as context, what makes Confucianism more empirical than empiricism—that is, what makes Confucianism a *radical* empiricism—is the fact that it respects the uniqueness of the particular and the need for a generative wisdom that takes this uniqueness into account in anticipating a productive future. Rather than advancing universal principles and assuming a taxonomy of natural kinds grounded in some notion of strict identity, Confucianism proceeds from always provisional generalizations made from those *particular* historical instances of successful living, the specific events recounted in the narrative of Confucius himself being a case in point.[28]

HOLISM AND ITS IMPLICATIONS FOR HARMONY AND ORDER

With a clearer understanding of the holistic assumptions of both Dewey and Confucianism, I now turn to a reflection on how harmony and order are to be understood within this interpretive context and how the connotations of these terms might differ from and need to be reconstructed out of a more familiar understanding of order. The holistic, vital cosmology embraced by early Confucianism and Dewey's pragmatism has many implications for the way in which the emergence of an always provisional, prospective order in the continuing human experience is to be conceived. For example, such a dynamic holism entails a doctrine of intrinsic relations that makes each particular event one of a kind, precluding the possibility of strict identity and replication. Thus "kinds" are defined analogically rather than by appeal to some self-same essential characteristic. Causality is not linear or separable from effect but is rather defused within and throughout such intrinsic relations, where each event is made possible by the always changing pattern of interdependent relations. Creativity is not wholesale but retail, expressed through the productive growth in these same relations. Because of the uniqueness of events, the emergent order entails the spontaneous emergence of novelty and hence is not reducible to antecedent causes. Creativity, and its expression in the human case as consummatory imagination, is radically situated as growth in relations.

For Dewey and Confucianism, harmony is melioristic: the continuing qualitative growth in relations that allows for the full accommodation of all of the participating elements in the totality of the effect. Such growth is achieved through *ars contextualis*—the art of mutual accommodation in the patterns of deference that constitute a thriving family and a flourishing community. It is the achievement of an aesthetic harmony that provides felt experience in our evolving humanity with its consummatory quality. Such harmony is the product of a situated creativity expressed as enhanced significance rather than as the necessary unfolding of some far-off and ultimate origin, as new ideas rather than as the instantiation of antecedent idealities.

It is for this reason that, in this holistic cosmology, harmony as agreement or conformity—a sense of harmony where the standard for conformity is antecedent, external, and privileged—has little relevance.[29] I would argue that in the Deweyan and Confucian aestheticisms, harmony cannot be understood in the sense of conformity because it precludes the optimal expression of the unique elements that come together to constitute an always particular experience of harmony.

Dewey describes the attainment of a participatory harmony out of tension and conflict in the following abstract terms:

> Equilibrium comes about not mechanically and inertly but out of, and because of, tension.... Form is arrived at whenever a stable, even though moving, equilibrium is reached. Changes interlock and sustain one another. Wherever there is this coherence there is endurance. Order is not imposed from with out but is made out of the relations of harmonious interactions that energies bear to one another.[30]

Dewey goes on to describe the attainment of an inner harmony within the human experience itself, distinguishing it from mere pleasure as a consummatory transformation of the conditions of life itself:

> Pleasures may come about through chance contact and stimulation; such pleasures are not to be despised in a world full of pain. But happiness and delight are a different sort of thing. They come to be through a fulfillment that reaches to the depths of our being— one that is an adjustment of our whole being with the conditions of existence. In the process of living, attainment of a period of equilibrium is at the same time the initiation of a new relation to the environment, one that brings with it potency of new adjustments to

be made through struggle. The time of consummation is also one of beginning anew.[31]

For Dewey, harmony is not achieved at the expense of particularity but because of the balance and equilibrium achieved among the various elements that constitute the harmony:

> There is an old formula for beauty in nature and art: Unity in variety. Everything depends upon how the preposition "in" is understood.... The formula has meaning only when its terms are understood to concern a relation of energies.... The "one" of the formula is the realization through interacting parts of their respective energies. The "many" is the manifestation of the defined individualizations due to opposed forces that finally sustain a balance.[32]

For Confucianism, it is quite specifically the family that is the ultimate source and indispensable ground of an achieved harmony or propriety (*li* 禮) in all of our roles and relations. As it states in *The Analects*: "Achieving harmony (*he*) is the most valuable function of observing propriety in our roles and relations (*li*). In the ways of the Former Kings, this achievement of harmony by observing propriety in our roles and relations made them elegant, and was a guiding standard in all things large and small. But when things are not going well, to realize harmony just for its own sake without regulating the situation through observing propriety in our roles and relations will not work."[33]

Confucian role ethics so understood exhorts the human being to aspire to that quality of conduct that makes relations stronger and thicker and more enduring. Without being properly situated within these roles and relations, actions are meaningless or worse. That is, a "harmony" that is effected by simply imposing external constraints as a means of enforcing order—the external imposition of laws, edicts, principles, or rules—is dehumanizing to the degree that it precludes personal participation and confirmation.

In a holistic culture, the continuing rhythm, tempo, and musicality of an achieved aesthetic harmony do the work of teleology, where the always improvisational satisfaction among unique particulars is both the process and the goal itself, an understanding of order that precludes as it does any dualistic divide between cause and effect or means and end and eschews any notion of a final vocabulary. To live intelligently is to focus one's attention upon the familiar affairs of the day and in so doing to bring lived roles and relations into meaningful resolution, distinguishing harmony from discord, musicality from concatenation, equilibrium and homeostasis from conflict.

HOLISM AND ITS IMPLICATIONS FOR SPONTANEITY AND REVOLT

A problem seems to attend our common yet paradoxical use of the word *spontaneity*. As Brian Bruya has observed:

> We can see the inception of what I call the Paradox of Spontaneity in the interpretations that Epicurus and Chrysippus give to natural human action. The Aristotelian internal teleology of natural objects, including humans, according to Chrysippus, allows us to say that all movement is spontaneous, that is, that there is no unmoved mover outside the natural order. For Epicurus, however, humans must remove themselves from the determinacy of nature and choose freely, thus spontaneously. This paradox—that movement is spontaneous as determined and spontaneous as free—persists even into present everyday speech (e.g. we may say that grass grows spontaneously, or that so-and-so spontaneously broke into song). Although the Greeks did not have an equivalent for "spontaneity" as a term of art, they clearly wrestled with this tension and were the foundation of all later thought on the subject.[34]

Synonyms for *spontaneous* are on one side of the paradox *impulsive*, which suggests randomness and disconnect, and on the other side *automatic*, referencing something that is "predetermined." Both meanings preclude any interesting understanding of the role of agency. The term *spontaneity*, rather than referencing a cultivated virtuosity, has come to connote either random, uncaused behavior or predetermined, automatic behavior. This development is closely related to the prominence of a *creatio ex nihilo* notion of creativity that reduces spontaneity to either an agent's impulsive and inexplicable action or to a preprogrammed design over which the agent has no control. Rather than suggesting a cultivated responsiveness on the part of the agent, spontaneity thus conceived refers instead to actions that are unconstrained and unstudied in manner. The spontaneous emergence of novelty on such a reading, far from being virtuosic, would not seem to entail any personal cultivation at all.

Dewey offers us a fundamentally different understanding of spontaneity as a collaboration between the meaning of past experience and the present stimulus that triggers its bursting forth in novel expression:

> Each of us assimilates into himself something of the values and meanings contained in past experiences.... Some occasion, be it what it may, stirs the personality that has been thus formed. Then comes the need for expression. What is expressed will be neither the past events that have exercised their shaping influence nor yet the literal existing occasion. It will be, in the degree of its spontaneity, an intimate union of the features of present existence with the values that past experience have incorporated in personality. Immediacy and individuality, the traits that mark concrete existence, come from the present occasion; meaning, substance, content, from what is embedded in the self from the past.[35]

Dewey rejects out of hand the play theory of art as a freedom "found only when personal activity is liberated from control by objective factors."[36] Spontaneity, far from standing in contrast to or in opposition to work and order, is the serious engagement with objective conditions to achieve a desired result: "The spontaneity of art is not one of opposition to anything, but marks complete absorption in an orderly development. This absorption is characteristic of esthetic experience; but it is an ideal for all experience."[37]

If we consult a Chinese–English dictionary, one equivalent given for *spontaneity* is *ziran* 自然. When this term is applied to human agency, the reflexive "self-" (*zi* 自) aspect of *ziran* has to be understood as persons in their contextualizing roles and relationships rather than as some separate and discrete self. Hence *ziran* is always collaborative. In fact, *ziran* is the uninhibited virtuosity that defines the character and conduct of *Zhuangzi*'s many craftsmen and enlightened exemplars as they interact efficaciously with their mediums and their environments:

> Butcher Ding was in the process of cutting up an ox for Lord Hui of Liang. Wherever hand came in contact, wherever shoulder leaned in, wherever foot trod down,

wherever knee exerted pressure—the twack of meat separating from bone would echo and the twing of the knife's blade would ring out, with every stroke singing a perfect note consonant with "The Dance of the Mulberry Grove" and the "Jingshou Chorale."

"Magnificent!" exclaimed Lord Hui. "That technique could be raised to such heights!"

Butcher Ding laid his knife aside and replied, "What I really aspire to is an insight into the way of things (*dao*) that takes one far beyond any mere skill."[38]

Spontaneity is the hard-won freedom of action captured in the description of an evolving virtuosity in the life of Confucius himself: "The Master said: 'From fifteen, my heart-and-mind was set upon learning; from thirty I took my stance; from forty I was no longer doubtful; from fifty I realized the propensities of *tian*; from sixty my ear was attuned; from seventy I could give my heart-and-mind free rein without overstepping the boundaries.'"[39]

To the extent that revolution is taken to mean one complete turn—a change so dramatic and far-reaching that it constitutes a new beginning—it again has little relevance for the holistic thinking of Dewey and Confucianism, and Daoism as well for that matter. James Carse makes a distinction between finite and infinite games that might be useful in distinguishing between two different ways of thinking about beginnings and endings in the human experience. A finite game is played between two opponents according to a given set of rules within a finite amount of time, with the objective of each of the players being to win. One chess game ends and another begins by coming back to the starting point. Or perhaps the opponents will try a different game altogether, introducing an even more dramatic new beginning.

An infinite game is different. The relationship between a parent and a child entails a different sense of agency that is not restricted by set rules or a given time frame, and the objective, far from being to win, is to strengthen the relationship in order to continue the play. It is this sense of infinite games that is a more relevant notion of progress in the holistic cosmologies of Dewey, Confucianism, and Daoism. Progress, simply put, is growth in relationships—a growth that is ultimately dependent upon personal cultivation and an educated imagination.

CHAPTER 2

What's Wrong with John Dewey and Confucianism and What's Right with Lao Tzu according to Eugen Rosenstock-Huessy

Wayne Cristaudo

> "Our modern World Society is as totalitarian in its way as Confucian China was."
> —**Eugen Rosenstock-Huessy**

Perhaps the strangest thing in "a strange book by a strange author," as Martin Marty once called Rosenstock-Huessy's book of 1946, *The Christian Future or the Modern Mind Outrun*, was the heading "Our Invasion by China." In another geopolitical climate that might sound like an alarmist reaction to China's economic rise, its cut-price exports and political friendships—but between 1940 and 1945, the years to which Rosenstock-Huessy explicitly refers in this section, China was neither exporting much nor invading anyone, let alone America. It was fighting for its very survival, engaging in what Rosenstock-Huessy calls "her first enthusiastic national war."[2] The United States, on the other hand, had been dragged suddenly and unexpectedly into the Second World War, a war it thought it could avoid. The rhetorical bite of Rosenstock-Huessy's formulation was largely being directed at North American pacifists, who he believed were living in a fool's paradise. They had, he says, "tried to make the upholstery of the suburb so thick that the rumbling of war and revolution could not be heard through the velvet curtains, the carpets and rugs of progressive education."[3] But it was not simply pacifism per se any more than it was China per se that was the object of Rosenstock-Huessy's critique. Rather it was a particular cast of mind that Rosenstock-Huessy was attacking, a cast of mind that he labeled as Chinese because he believed it had originally been developed in China by Confucius. And it was now becoming a dominant cast of mind in the West through the influence of people like John Dewey and the movement of pragmatism.

Of course, Dewey had gone to China to instruct what then seemed like a fledgling democracy about social development, and his ideas were very popular with a milieu of intellectuals in China. But, as with Confucius, his writings were proscribed for a long time by the Communist Party. As for Confucius, he has not only been reassessed in China, but he has become the symbolic figure of orthodoxy for a party wanting to modernize peacefully. That this is not unproblematic is all too conspicuous in the rather strange event in 2011 when a newly erected statue of Confucius in Tiananmen Square simply disappeared.

In any case, it is not surprising that, along with Confucius, Dewey's star appears to be shining again in China. That was certainly the hope behind the coupling of Confucius and Dewey undertaken at the turn of the millennium by David Hall and Roger Ames in their book *The Democracy of the Dead: Dewey, Confucius, and the Hope for Democratic China*, which was an attempt to find a more enduring and productive path to democracy for China than the individualistic rights-based democracy so often trumpeted by American liberals.[4]

For Rosenstock-Huessy, who escapes any classification along liberal or conservative lines, Dewey and Confucius do indeed form a natural alliance.[5] As Rosenstock-Huessy saw it, Dewey was "the patron saint of progressive education, has emancipated the mentality of our suburbs from any subservience to Church or State. He has become the Confucius, the educational sage, of the Western World."[6] "Master Confucius Dewey," he says, "well might be our social saint."[7] Further, for Rosenstock-Huessy, it has not been the overt trade in ideas and commodities that has introduced Chinese ideas into America, but pragmatism.[8]

To understand the thrust of Rosenstock-Huessy's attack, it is worth quoting the following passage that lays out what he thinks is fundamentally wrong with Dewey and Confucius:

Confucius and Dewey are very wise, very old, very kind and patient, very sure of being on the inside. Hence, prudence, justice, temperance, industry, self-control are their virtues. Cannot the murderer be improved, the wicked be enlightened, the wars abolished? And revolutions can be avoided. Since man, as they are convinced, can be sure of this, he need not get excited when catastrophes do befall him. They need not make us unhappy since they need not to be. We may overlook them by anticipating the certainty of being inside everything. To them progress is ingress. They mean by it the constant entrance of more and more people into the inside order already familiar. Progress is not the revolutionary beginning of a tradition hitherto unknown but the extension of known qualities. Progress is painless and not the heartrending conflict of previous progress now hardened into tradition and future tradition initiated as progress. To the man who believes that we are creatures, our own accomplishments of yesterday stand in the way of the next accomplishment because the old traditions are sanctified by sacrifices made. But Dewey says that intelligence for millions of years was led astray and has now found itself. Within the frame of millions of years, it is childish to weep over any loss of country, loyalty, love of old; we may keep smiling and this indeed is what we are told.[9]

The last sentence is certainly polemical, and while that sentence may not be completely fair, the general point is a powerful one, and one I think we might well want to extend beyond a mere discussion of Confucius and Dewey but to a paradigm based upon a faith that is so pervasive today that one might at first wonder why Rosenstock-Huessy has a problem. Is it not good to want harmony? Is it not good to want stability—not good to retain order and bring the disorderly into a greater order? And when he speaks of "our stress on adjustment to environment, the avoidance of conflict, the pragmatic value of truth, our concern for practical success" (which he emphasizes are all "reminiscent of [Confucius's] China"), bafflement might well be the first response to his comment that they all conspire to form a totalitarian mind-set.[10]

To get some idea of his problem, let me quote another passage from the book. This time, though, it expresses what he favors:

> The Cross explains war and revolution and decay and disintegration and explains why some sacrifice must bridge the gaps which man's abuse of his freedom always rips open.

> Free men must shift their allegiance from solidarity and functioning "inside," to rebellion, to reverence, to sacrifice, according to the evils which have to be resisted most urgently.[11]

The first and obvious point to mention about the quote is that it shows Rosenstock-Huessy nailing his flag to the mast of Christianity. But let us add immediately that just as he has identified what I will call the core elements of the paradigm he is critiquing, his appeal to Christianity is to a core that Christians should all recognize but that they all too frequently neglect in favor of what for Rosenstock-Huessy is but the mythical, legendary, philosophical, and theological aspect of their religion (what he saw as the childish and overly abstract bits, to put it bluntly), as opposed to what is central in the *how and what* of Christian world making through the church and what he sees as its offspring and, to use a term that might no longer seem appropriate or accurate, the Christian nations.[12]

Rosenstock-Huessy's Christianity rests almost entirely on the insight that the cross and what it represents, namely suffering, sacrifice, and love, are the primary means by which the human race purchases a better future, one in which certain fundamental problems of humankind are solved once and for all, and thus the effect of their cumulative institutional integration enables a greater degree of human freedom than could have been attained without them. More specifically, he argues, the progress that the human race has achieved has mainly been acquired through the church's capacity to unite Western Europe in the first millennium and through the complete reconfiguration of human association and potentialities that transpired throughout the second millennium through the great and total revolutions—specifically the Papal Revolution or Investiture Conflict, the Italian Revolution or Renaissance, the German Revolution or Reformation, the English Revolution, and the French Revolution culminating in the world wars, whose revolutionary spill-outs were the Russian Revolution and Chinese Revolution. This hybrid of "progress" *and* ever greater catastrophe enforced upon humanity a choice between a common global future and no future at all. Although the Russian, Chinese, and French revolutions were either anti-Christian (the French Revolution) or explicitly materialist and atheist (the Russian and Chinese revolutions),[13] they still aspired to the same messianic ends and universalism as the overtly Christian revolutions that preceded them. In that sense, he believes that their hostility to the church was to the more local circumstances of the church's failures—in France the Catholic Church's refusal to reform itself and its reactionary and parasitical alliance with the French monarchy, and in Russia the insistence of the Orthodox Church's decision to "opt out of history," a strategy that left Russia's social classes deeply riven. (In China one may point to the church's association with colonialism.) The revolutionaries of France and Russia had completely failed to understand the meaning of the Holy Spirit, that spirit that refuses to rest in institutions that do not live by the commandment to love. And for Rosenstock-Huessy, the volatile, indeed explosive kernel of Christianity was that in preaching neighborly love, its people became more ferociously hostile to institutions that offered no redemptive future because they were simply self-serving for particular groups or classes. Hence he can make such claims as, "Through Communism, the Chinese for the first time have entered the Western idea of one history.... And therefore at this moment, there is no longer a specific pagan pre-Christian Chinese history,"[14] and "The non-Christian side of French Jacobinism is really its most Christian side. It offers to the Jew a common meeting ground on the basis of humanity, of humanism."[15]

For Rosenstock-Huessy, the Holy Spirit as a power that requires Western peoples to assess the Christian character of their own institutions stands in the closest relationship to Christianity's unique insight into future making—that is, its realization "that death precedes birth, that birth is the fruit of death, and that the soul is precisely this power of transforming an end into a beginning by obeying a new name."[16] Like all revolutionary insights, the insight at the heart of Christianity radically overturns the "natural" or "normal/obvious/commonsensical way" of seeing things; in particular it inverts the usual relationship between birth and death so that one must die into new forms of life by leaving behind what is spiritually dead. This is also related to another essential feature of Christendom: that redemption involves resurrection. And for Rosenstock-Huessy, it is no accident that where Christianity takes root, it revives and resurrects forms of life and institutions thought dead or spent and it does so because this is at the core of its understanding of its own redemptive purpose. One need think only of how the Greek legacies of democracy, the arts, and the academy were all not only revived but reconstituted on Christian soil. These "resurrective" acts were themselves closely allied to passionate and sacrificial acts of faith in a better future. From the little I have said here, I hope it is evident that Rosenstock-Huessy's interpretation of Christianity is primarily social and radical. He is not interested, for example, in what happens to the soul after death; he is not interested in another world but in how the spirit informs the creative and redemptive acts through which we make our future. He is, one might say, a Christian realist. That is, he is interested in what kind of reality Christianity generates.

In *The Christian Future*, while he attacks what is essentially the "liberal progressive" view of social life for what he thinks are its smug and unself-critical assumptions, emphases, and—not to put too fine a point on it—totalitarian approaches to truth, he says in passing of his Christianity: "We have conceded to them [liberal progressives] the obsolescence of the previous Christian 'set-up.'"[17] The "we" is a rather small group: Rosenstock-Huessy and Christians who are conscious of having moved through the three different ages of the church. That idea of the three ages of the church was originally expressed by Joachim of Flore and in more modern times by Schelling, and it also was a central part of the analysis of Rosenstock-Huessy's Jewish philosopher friend and dialogical disputant Franz Rosenzweig in *The Star of Redemption*. In the case of Schelling, Rosenzweig, and Rosenstock-Huessy, the Christian faith in the West is seen as having evolved into three different formations: the Roman or Petrine Church, with its visible spiritual signs pointing to a world other than this and with a mission of uniting people through these common signs and symbols; Protestant or Pauline Christianity, which renounces the renunciative and centralized Petrine Church and spiritualizes this world and which no longer needs visible signs of God's presence but emphasizes the inner faith of the believer and thus allows much greater freedom for a Christian conscience; and, finally, what Schelling called the third age of the church or the Johannine Church, the type of Christianity I have just mentioned.

Significantly, the Johannine Church is largely made up of people who have been stamped by the other ages of the church but who no longer even think of themselves as Christian because their faith in universal solidarity no longer requires any specific institutional configuration or any specifically Christian symbols. The danger with this age of the church is that its "members" lack not only visible signs but even a sense of the fundamental language of Christianity, so that they are rarely aware of their own heritage. Thus it is that Rosenstock-Huessy says he is recasting the Christian truth—for Christianity, according to Rosenstock-Huessy, must ever reinvent itself, breathe life into new names, and recognize that old names may have exhausted their power. To make his point he even cites Hans Urs von Balthasar's remark that "The word 'God' is so spent that we do not intend to haggle with Nietzsche on its behalf."[18] In this respect, when Rosenstock-Huessy wrote *The Christian Future or The Modern Mind Outrun* in 1946, he made it very clear why he had written a book with such a title, which seemed so contrary to the obvious fact that Christianity, in the West at least, very much looked as if it did not have a future. He was writing to show people what "they had almost lost,"[19] a particular faith that frequently was embraced in purely legendary or mythical form but that in actuality was an orientation toward life whose fruits, for Rosenstock-Huessy, were now visible not in the church but in the institutions that came out of the great revolutions that took place originally on Christian soil. Thus in *The Christian Future*, Rosenstock-Huessy puts forth the case for the Johannine form of Christianity, a Christianity of "hope." And he calls it a "listening Church, [which] will have to unburden the older modes of worship by assembling the faithful to live out their hopes through working and suffering together in unlabelled, undenominational groups, thereby [ready] to wait and listen for the in break of a new consolation which shall redeem modern life from its curse of disintegration and mechanization."[20] Note in passing also that the battle is against disintegration and mechanization. I say this because he sees the Dewey/Confucius nexus and indeed the entire mistaken direction of modernity being bound up with the folly of its faith in mechanization/functionalism—a faith that ultimately will not only not stop the disintegration but will create even more havoc than the explosions it seeks to stabilize.

Rosenstock-Huessy's faith, although reached before the Great War, was consolidated and given its particular line of direction by that war. For it was in that war that he was sufficiently traumatized *to see* the great waves of history that had led him into the midst of the mass slaughter and carnage that he experienced as a captain at Verdun. He was driven to write *out of the experience of war* to try and help prevent future war. In this one might even say he was working toward the same goals as Dewey and Confucius. But his faith in the *eschaton* is a messianic type. That is to say, the harmony for which he strives is that in which the full birthright of man—his divine nature—is fully realized. Thus any kind of harmony that would involve the truncation of the powers of the soul of any group or that would seek to provide order and harmony of materials that are not fully formed (biblically speaking completely in God's image) and thus spontaneously harmonious is delusional. (The affinity with Goethe is obvious—and his friend Rosenzweig said that Goethe was the father of the Johannine age of the church.) Certainly the idea that our divinity will be attained primarily through education, sagacious rule, and virtuous leadership is, for Rosenstock-Huessy, a complete fantasy. (Recall the original plight of Faust: that all his learning has left him bereft of the experience required to make him truly become who he is.)

We mentioned earlier that Rosenstock-Huessy's vision of history was born in the trenches of the Great War. We should not forget that prior to the Great War, Europe had rarely if ever had such a long period of sustained peace. The balance of power, the proliferation of international law, and international peace treaties throughout the latter part of the nineteenth century and into the early twentieth century seemed to be enough to provide the peace, but they did not. On the contrary, the forces of nationalism, imperialism, and military and economic growth—even the spread of democracy and international trade—conspired not to provide peace but to plunge humankind into the bloodiest of centuries. The peace had been poorly used. That peace is something that must be wisely used, that it must be paid for, and that war is the inevitable result of misspent peace flies in the face of the entire liberal faith in humanity's natural goodness and innocence. For liberalism generally assumes—and this is clearly laid out in Locke's *Second Treatise*—that people are inherently cooperative and reasonable and thus that cooperation and reason combined with law and representation will create peace.[21] (This faith is also behind the widely repeated, and seductive, "myth" that democracies will not go to war against each other.)

Liberalism, especially in the United States, is often criticized for an exaggerated faith in individuals and their rights. And thus, as in Dewey, the argument is made that a more collectivist or communitarian ethos is required to provide social stability and harmony. That, of course, is the argument made by David Hall and Roger Ames, but for Rosenstock-Huessy, the problem is not adequately cast if it is seen as a contrast between individualism and communitarianism, or for that matter democracy versus bureaucracy. Rather the argument that has to be made is the relative importance of sacrifice, suffering, and love and the kinds of virtues laid out by Dewey and Confucius. Of Dewey, he writes:

The ideal society is conceived by him and his followers—and these practically are the teachers of America at this moment—as a

scientific
democratic
depersonalizing
cooperative

functional mechanism, in which all the individuals who agree to it are held together by what they call social intelligence.[22]

He adds that while Dewey is himself a Christian liberal, his followers believe:

1. Society is God and otherwise there is no god who sends us into the world by calling us by our names.

2. Therefore, human speech is merely a tool, not an inspiration; a set of words, not a baptism of fire.

3. Society includes all men regardless of their evil character. Everybody can be educated or re-educated. The body politic needs no self-purification.

4. The ipse dixit of authority is always out of place. Conflicts can be solved by discussions between equals.[23]

As for Confucius, he finds a very similar set of priorities; Confucius "was impersonal, functional, silent about God, un-emphatic, democratic in education."[24]

For Rosenstock-Huessy, Dewey and Confucius both appeal to a politics of peace, a politics that comes after war. Indeed, he sees (wrongly, as I clarify below) a deep similarity in the fact that both men belong to ages and locations that seduced them into believing that the chaos of life had been tamed. Dewey, he says, was formed by the America of 1890, an isolationist America devoted to the "cogs on the wheel of industry,"[25] while Confucius "came when the China of the hundred tribes had been welded into one empire. Confucius could hold that politics was education, since by that time everybody was inside the Chinese wall of one empire."[26] According to Rosenstock-Huessy, in so far as Confucianism and Deweyism promise a recipe for peace and good sense, they are very appealing. And in so far as they both have so much chaos behind them, they represent "the needs" of their age,[27] but those needs that come from a desire to be free from chaos. And the illusion that chaos is contained—Rosenstock-Huessy sees the Chinese version in "birds in cages. Waterfalls in gardens … dragons on boudoir tables"[28] and the American variant in Yellowstone Park [29]—comes after forces have indeed been tamed and peace looks as certain as it is welcome. Then wisdom, duty, experiment, and discussion all stand very tall indeed. Then it looks as if the world may well conform to the essentials of Deweyism and Confucianism: "one scientific silently functioning all inclusive cooperative impersonal painless order, an order in which nothing vital has to be settled by force; nothing lunatic can ever befall whole nations, no personal decision must save the world from ruin."[30]

Rosenstock-Huessy's argument against the danger of idealism is a powerful one, but his comments on Confucius are historically inaccurate. As Bill Ratliff has written to me, Confucius was looking back to the "largely mythical unity and high virtues of the early Zhou dynasty rulers, especially the Duke of Zhou. But RH's comments on 'one empire' 'inside the Chinese wall' clearly imply the unity enforced after 221 BC by Qin Shihuang after a very, very nasty period of violence and warring states."[31] It is also a dubious claim that Dewey never changed his thinking to take into account the catastrophes of the world wars. Indeed, I think Rosenstock-Huessy's mistake here is not that he draws attention to the failure to connect the forces of strife with those of peace but that he considers idealism as the projection of a peace already experienced instead of as what it clearly was—for example, in Plato, the projection of a desired state of affairs that would be completely different than what was the case. On the other hand, Rosenstock-Huessy's experiences as a soldier who fought at Verdun also left an indelible impression upon him about the meaning of war within life.

In any idealism, what is lost is a consolidation of *real* possibilities: where we have been, how we have traveled, what we are, and what we may be. Thus, against this view of politics as education, planning, wise counsel, and intelligent functioning—qualities that have an important place at the right time within appropriate limits—Rosenstock-Huessy in *Out of Revolution* writes of sacrifice and politics:

> When and where we love or fear, we are willing to pay. We are willing to spend money, or in more serious turmoil to sacrifice some parts of our own nature, and to consecrate others. We are ready to forget certain temptations, and to give free rein to others. Thus, our energies flow into new channels each time that our hearts leap. And each leap of our hearts remakes our bodies, our habits, and our institutions. Since any heart that has the privilege of loving is willing to suffer for its love, our social customs are the fruit of these sufferings which reshape our ways of life. The Body Politic as well as the cellular body is the reward of the sacrifices which our heart has paid for its privilege to love.[32]

Earlier I spoke of war as a consequence of misspent peace. A persistent problem that still confronts human societies is that they often do not know that they are misspending the peace. People are invariably too busy enjoying its fruits to notice that they are contributing to future social breakdowns that will escalate into war. To look to the future with the eye of peace is to be half-blind. The fact is that almost all of the institutions that we in the West now look to for preserving the peace and expanding human capacities came out of war and revolution—whether it be the nation-state, liberal democracy, private property, or even what Harold Berman has called, in the second volume of his *Law and Revolution*, "the twin legal innovations introduced by the Russian Revolution": "the enormous enhancement of the social and economic role of the state and the parallel enhancement of the parental role of law."[33]

In his studies of revolution and war, Rosenstock-Huessy is brilliant at diagnosing the various sources and triggers of explosions, but it is also a sad fact that the causes of war are easier to see after its outbreak. Having dedicated himself to looking for where wars and revolutions came from, he also could not deny that in the case of Europe at least, whatever major institutional progress had happened had been paid for by sacrifice. Now not all sacrifice requires war, but all wars require sacrifice. And they do so whether the sacrificed are willing or not. Peace is a different matter. First there is a problem of people being prepared to make the necessary sacrifices for the future—peace is often pleasant, and who wants to interrupt their peace for a future that is far away, and why should it cease to be as peaceful as it is now unless people are just crazy? (Peacetime thinking likes common sense.) Of course, not all flourish in peace, but by and large, those who suffer in peace—who dwell in the mental homes and prisons and sites of addiction and loneliness—do so privately, and their disintegration can largely be left to them and their families, social workers, and other state employees.

But there is another, thornier problem about peace, and that is that even if some are prepared to sacrifice themselves to a better future, there is the question of where and to what should they make their sacrifice. For how attuned are those who are prepared to make a sacrifice? Are they attuned to the disintegration happening now or do they tend to look for where the last outbreak of war occurred? My question implies my answer. A good example of what I am talking about can be seen from the radical liberal paradigm that was erected by what we may broadly call the '68 generation. That paradigm, which has become ubiquitous in universities throughout the world and has fed into so much social and political discussion, policy, and legislation in the West, is essentially devoted to eliminating social fascism, all vestiges of racism, and prejudice against and deprivation of the rights of members of certain classes, races, genders, sexual groups, and so on. That is to say, a great deal of time and energy has gone into ensuring that the Second World War will not happen again. And it won't. If there is a Third World War, it will no more be caused by social fascism than the First World War or the Thirty Years War were. Having said that, one need only cast one's mind to the imperial forces in collision at the time of the First World War to see how precarious the peace is in 2014. Yes, Germany and France have been settled, though none would have thought that the blowback of the end of the Austro-Hungarian Empire would take place near the end of the century in the Balkans, the very area that had ignited the Great War. Japan *seems* no longer to harbor imperial aspirations, but are the Chinese convinced of this? Indeed, have Japan and China really made peace? The Ottoman Empire and its periphery is a problem still not remotely solved, a gaping maw of political and social instability with Israel a nation permanently ready for or at war with its neighbors. Turkey is caught between the West and Islamic visions of society, and it is hard not to see the AKP as harboring hegemonic aspirations in the region. In this respect, and turning to the periphery of what used to be that empire, it finds itself in a dangerous balance of power with the Saudis, Iran, and, most unpredictably, Al Qaeda. The Russian Empire received a boost of life by the Soviets, but now it is a tottering, humiliated giant mourning its losses, with a number of ex-Soviet states caught up in their own domestic chaos fueled by corruption, resources, and Islamism. The United States, with everything before it in 1914, still shows itself oscillating between hegemony and isolationism and seems to be decisively unsuited to the former role, as its post–World War II history of failed interventions illustrates. In the aftermath of the British Empire, India and Pakistan are ever a heartbeat from an explosive outbreak of fresh hostilities. China now seems well on the way to being a superpower. But the Cultural Revolution, historically speaking, is a relatively fresh wound. And accelerated growth, rising aspirations, and high degrees of inequality are always a volatile mix. China also has twenty million Muslims, and the more than eight million Uyghur in the Xinjiang area are a source of permanent instability. The problem of Tibet is likewise unsolved, as are ongoing claims to Taiwan, as well as territorial disputes with its neighbors, the dispute with Japan over the Diaoyu Islands being the most conspicuous and potentially catastrophic one. And the imperial desertion of Africa has left almost no stability on the continent. It is a political void that can only aggravate tensions between greater powers.

In sum, The peace has not been well spent. The same faith in economic progress, arms races, and nationalism is there—and I cannot help myself: people think religious people are superstitious! If I have moved from Rosenstock-Huessy, I have simply taken his thought into our time. And if I may now return to Dewey and Confucius, I do so to emphasize that Rosenstock-Huessy's insights into the relationship between sacrifice and suffering, and his critique of a way of thinking that promises peace through ingression, are more desperately in need of being listened to today than ever.

In the case of the resurgence of Confucius, the reason for that resurgence is so compelling that one must be really vigilant to resist the temptation. The reasoning of Dewey and Confucius—and one sees it just as strongly in Aristotle, who shares their decency, good sense, and desire for harmony—is one that relies on the inheritance of an achieved order. In the lengthy citation about Dewey and Confucius above, one sentence contains the key of the entire mistake: "Progress is not the revolutionary beginning of a tradition hitherto unknown but the extension of known qualities." Of course, when Dewey went to China to show it a better future, he came bringing all the fruits of religion and war that he had inherited in the West, but he spoke as if they were the pure fruits of reason, as if brains and discussion had been the vital ingredient in their existence. In this respect he was indeed a sage much like Confucius, and the temptation was to think that a sage could save the nation. He didn't and he couldn't any more than Confucius did or could have. Sages already have an ideal mental order they wish to protect, but that means that those who are crushed, negated, or driven mad by that order are ignored; their enslavement is preordained as part of the order of heaven, or the way things are. Revolutions and wars, however, are not the products of deranged and deformed brains that fail to conform to the rational as such but are the results of often unseen or unnoticed volatile toxicities that too often escape the purview of all but those who are suffering from them.

While Rosenstock-Huessy had been thrown into the hell of the First World War, Dewey was remote enough from it to remain essentially untouched by it. Of course, he discusses war and takes a position on wars, but the very fact that he believed that war could be outlawed is a most revealing sign of the difference between his way of thinking and his *faith* and that of Rosenstock-Huessy.[34] Although Rosenstock-Huessy made speech the cornerstone of his thinking—he and Rosenzweig both called their own paradigm speech thinking, and he can be correctly classified as a dialogical thinker—he does not exaggerate what dialogue can achieve. For speech is often a declaration of what cannot be compromised, of truths that must be fought for, of war:

> These are the facts—some of them—which upset the pragmatic universe: 1. No youthful nation including America has ever settled vital questions by discussion. St. Augustine said that discussion is for those questions which do not deal with the necessary. We can discuss in a democracy, everything. But we cannot discuss anything with those who reject discussion. 2. The German youth threw out all teaching authority wholesale, and followed an untaught Fuehrer. It denied all intellectual authority which Dewey and Teachers College take for granted. 3. Bolshevism has destroyed millions of victims. 4. Hitler has murdered millions of Jews. 5. America cannot find a rational solution of the Negro question after one hundred years of search. 6. Our armies had to gain the initiative by bold decisions.[35]

Against the faith in education, discussion, cooperation, function, and painless order, Rosenstock-Huessy holds:

> All the insoluble cruxes like unhappy marriages, race strife, injustice, are not borne by reasoning but by the eternal combination of three irrational qualities: forbearing charity against the perpetrators, flaming

defense of the outraged victim, reverence for the inscrutable decree of providence. Our faith in forces greater than man's intelligence, a charity greater than any social intelligence ever warrants, and unbending hope in the victory over the worst fiend, animate those who by their personal decisions and sacrifices enable Confucius or us to cooperate and to live inside of some semblance of order.

Spiritual authority, sacrifice, creative exuberance, aye, ecstasy, sufferings, are creating the frame of reference "inside" of which Dewey's army of teachers alone can work.[36]

In sum, the view of order and harmony that Rosenstock-Huessy identifies as essentially common to Dewey and Confucius is one that fails to fathom sacrifice, suffering, and love as the creative conditions of the peace they inhabit. Dewey's and Confucius's failing is that they do not adequately comprehend not only the necessary sacrifices of those prepared to establish the peace, or determined to uphold a peace achieved, but those fiery energies of outrage and wrath, the extent of desperation of souls determined to overthrow what they see as the rotten buttresses of an existing order that ensconces suffering, humiliation, and unbearable limits.

I have said much about the West thus far, but I hope from what I have said already that it is clear that Rosenstock-Huessy does not have truck with any essentialist notions of the East and West. He is far more interested in the plights and institutions of peoples. And while each people has its own history, Rosenstock-Huessy is more interested in the commonality of our plights and the common search for solutions. Thus he argues in *The Christian Future* that now that we inhabit one world we are all heirs to four great traditional world-making streams—Christianity, Judaism, Buddhism, and Taoism—which, each for distinctive reasons, can assist us in invigorating modern life and thereby help save it from "the curse of disintegration and mechanization." He sums up what such an interpenetration for the social sciences will mean thus: "The immense material provided by the research of the social sciences is like the Old Testament of the World which waits to be read with the eyes of Buddha, to be listened to with the faith of the prophets, to be harmonized with the ease of Laotse and to be incarnated with the love of Christ."[37]

I have examined his discussion of Judaism in detail in my *Religion, Redemption and Revolution* and wish to emphasize here that in the case of the Buddha and Lao Tzu, Rosenstock-Huessy argues that their value today lies in their reactions to ancient problems that have returned to haunt modern men and women with a vengeance. In the case of the Buddha, Rosenstock-Huessy sees that he opened up a path to respond to the oppressiveness of the Hindu cast system and the ultrarelativism accompanying the Hindu proliferation of divinities. He believes the West must learn from the Buddha's insights and responses to the intense scope of violation that pervades social life. Today, the scale of exploitation and violation of the world and human beings through the intensification and acceleration of the modern mechanization of life is unprecedented. For Rosenstock-Huessy, the Buddha is a role model for how to extinguish desire in the massive process of the hyperobjectification of the world.

In the case of Lao Tzu, what intrigues Rosenstock-Huessy is the profundity of his response to the ultrafunctionalist approach that is as ubiquitous in the modern world as it was in ancient China. Furthermore, he sees that whereas the Buddha's dissolution of *maya* places him outside social and political life, Lao Tzu finds a way of dealing with the social and political from the inside. Just as Rosenstock-Huessy sees the twin totalitarian folds of pragmatism and Confucianism as attempts to cut off the very incalculability that is the accomplice of genuine vitality and creativity, he sees Lao Tzu's "genius" as the refusal to conform to the mechanized/functionalist paths laid out in a mental matrix of controlled duties and sagacious instrumentalities. Instead he attunes himself to a more primordial creative hum that one reaches not by functioning but by nonfunctioning, by choosing the unimportant, the space between things and words rather than the things and words themselves. The way of Lao Tzu is not, says Rosenstock-Huessy, "the straight highway"[38] —which is why he likes it so much, for real life is never a straight highway; at best one might travel the highway for a brief span—but the detour. The detour is often hard, but it is through that hardness that we may learn something that could never even be noticed on the highway. Amid all the mechanization and functioning, we are desperately in need of being reminded that the rhythms of life and the pathways of life will not all fit in the head of a sagacious educator. But we must endure in this world, and Lao Tzu, says Rosenstock-Huessy, teaches that the "Tao is the art to live with endurance on the way."[39] And instead of us all lining up to be slotted into the great universal machine that modern men and women are hell-bent on making to plan a more secure future, Lao Tzu gestures to us to enter into anonymity and vanishing, to enter into the "effortless centre of non-activity on which all things turn," so that we may not simply fit into the cog of society but "dance the universe."[40] Lao Tzu is the polar opposite of the modern mind with its plans and fixed points of certitude, its timetables and efficiencies. Just as Marx reminded liberals that producing is as much a part of being human as consuming, Lao Tzu is a constant reminder of the very qualities that compelling contributors to the machinelike world we are building, such as Confucius and Dewey, ignore; that is, the traceless, the nameless, the weightless—all those useless qualities that escaped the order of heaven as much as they escape the order of our instrumentalized, economically efficient but increasingly spiritually dead social cosmos.

Rosenstock-Huessy's friend Franz Rosenzweig held that the truth of revelation is the "secret" revealed in the Song of Songs: that love is as strong as death. To an important extent this is also the real core of Rosenstock-Huessy's faith and his fight. For him, the Dewey/Confucius/modern nexus neglects this, but Lao Tzu's way is a way that all, whether from the East or West, should follow. And the sooner our talk of East/West traditions—indeed our whole infatuation with spatial locations in a world that has so radically reconfigured time and commun(icat)ion—is dropped and we combine our energies to distinguish between the paths of the living and the dead, the better.

CHAPTER 3

Mohism, Standards, and Social Order

Donald Sturgeon

Alongside the early Confucians, the Mohists were one of the most active groups of thinkers and social reformers in the Warring States period of ancient China. Yet though Warring States sources cite both Mohist ideas and Mohist devotees as being highly influential at the time, the end of the Warring States period, and particularly the emphasis placed on Confucian ideology from the Han Dynasty onward, led to the anthology of Mohist writings known as the *Mozi* being largely neglected for many centuries.[1,2] Indeed, the majority of the *Mozi* as preserved today survived largely by historical accident, having been included in a collection of Daoist texts.[3] Decline in the importance of the text, combined with the loss of dedicated groups of Mohists who promoted the ideals discussed within it, contributed to centuries of textual corruption. Until the work of Chinese textual scholars in the eighteen and nineteenth centuries, the text was in such poor condition that some parts of it were effectively unreadable. Despite this, it seems clear that the Mohists contributed greatly to the philosophical discourse of their time, and though unsympathetic commentators have dismissed the philosophical depth of their ideas, others view them as the most philosophically important group of the early classical period.[4] Even so, it is worth remembering that the primary aim of the Mohists in their own time was practical restoration of order to a chaotic world rather than an exercise in abstract reasoning.[5]

In this paper I argue that although the Mohists in fact shared many values with the early Confucians, their rejection of both the concept and content of the rites (禮 *li*), and replacement of them with their conception of standards (法 *fa*), led them to come to a quite different view of a well-ordered society than the Confucians. Whereas Confucians placed much importance on following a traditionally effective system of rites, which could be supplemented or even overruled by invoking other important values, the Mohists aimed to produce a system of standards by which all actions could be judged, thus producing a single code that, unlike the rites, should always be followed to the letter. Although the Mohists were not interested in reforming the basic structure of society, they did attempt to bring inclusiveness and consistency to their code of conduct in many of their key principles; for them, consistency and conformity appear to be keys to a harmonious society. Though the Mohists do not argue that people are inherently equal as individuals, the same standards apply to a ruler or official as to anyone else—an idea that contrasts with early Confucianism.

COMMON BACKGROUND: REN (BENEVOLENCE) AND YI (RIGHTEOUSNESS)

Many doctrines presented in the *Mozi* can be viewed largely as reactions to the actions and established practices of both the *ru* (儒 ritualist; later Confucian) class and other "gentlemen" of the time; many of these practices were actively endorsed by Confucians but were rejected by the Mohists as wasteful and unjustified.[6] Despite this, the Mohists shared considerable agreement with the Confucians on a number of basic ethical matters.

Both the *Mozi* and the *Analects* appeal to the concept of benevolence (仁 *ren*) and to the "benevolent person" as idealized role models that all morally good people should aspire to emulate. The texts also agree on a number of exemplars of exceptional, praiseworthy individuals variously termed sages or sage-kings, such as Yao and Shun. They argue that morally good people should pursue and endorse what is right (義 *yi*) as well as oppose what is not right (不義 *bu yi*). Other important virtues that a morally good individual should embody include filial piety (孝 *xiao*) and respect for elders (悌 *ti*).[7] A common theme of Warring States debate was the importance of order (治 *zhi*) and devising ways to prevent its opposite, disorder or chaos (亂 *luan*); though this theme is much more central to the *Mozi* than to the *Analects*, it seems clear that both texts also endorse the value of order.[8]

Despite these initial similarities, the Mohists and Confucians were seen at the time as two influential but diametrically opposed groups of thinkers.[9] The two groups, though they might have been able to agree on certain broad aims of their respective social projects when expressed in general terms, such as the importance of promoting benevolence, righteousness, and order, differed fundamentally on the correct interpretation of these ideas.

RITES AND STANDARDS

Though Mohists and Confucians might have been able to agree on a few fundamentals, there are two crucial ideas upon which they disagree: the importance of *li* (禮 rites, ritual)[10] and the importance of *fa* (法 standards, laws) in ordering society. Most obviously, the *Analects* places enormous importance upon the concept of *li* and emphasizes that the morally exemplary person is one who conducts himself in accordance with these rites.[11] The *Analects* also contrasts governance by means of orders backed up by threat of punishment—which it says will make people "avoid punishment, but have no sense of shame"—with governance by means of virtue (德 *de*) and *li*—which will make them have shame and goodness.[12] The Mohists, on the other hand, are concerned with the Confucian *li* only insofar as they wish to condemn certain aspects of it, and they instead attach great importance to justifying and motivating people to follow the correct *fa*—motivation that the Mohists explicitly say should include reward and punishment.

Whatever further significance *li* has in the *Analects*, it seems clear that in one respect it is something like a code of conduct: it gives specific guidelines for exemplary action.[13] For example, when asked about filial piety, Confucius's answer is that "parents, when alive, should be served according to the *li*; when dead, buried according to the *li* and sacrificed to according to the *li*."[14] According to the *Analects*, *li* is also essential to avoiding various types of problems and disorder,[15] and is at least one means of ensuring order between ruler and ruled.[16]

Though there are various views on whether *li* is the most fundamental or important concept in the *Analects*—other suggested candidates include benevolence (*ren* 仁) and righteousness (*yi* 義)—it seems clear from the text that rites play a very important role. One passage of the text appears to explain the crucially important value of benevolence simply as adherence to the rites in all matters.[17] Yet other passages state that Confucius himself advocates going against certain rites,[18] and later Confucian texts note that there are moral conflicts in which disobeying the rites is very clearly the morally good thing to do.[19]

Since *li* is in some sense a code of conduct, much as with any scheme of codes or laws, interpretation must play a role in determining whether or not a particular rule applies in an actual concrete case.[20] One might be tempted to suggest that various other virtues or values important to the *Analects*, such as benevolence, might then be viewed as something like "interpretative principles" that govern how the *li* should be interpreted in individual cases and contexts—particularly when we are told things like "in applying *li*, harmony (和 *he*) is of the utmost."[21] But such an explanation tends to draw attention away from the fact that in some cases *ren* can simply override *li*, and it is sometimes the case that one should not follow the *li* because doing so would not be *ren*. So it is not the case that one should always follow the *li*; but neither is it the case that one can ignore it a large part of the time. This leaves us with a code of conduct that should normally be followed but to which there can be exceptions on other grounds. In other words, it has a considerable degree of flexibility and may be open to significantly differing interpretations in particular cases due to external factors.

Leaving aside the apparent tension between rites and benevolence, it at least appears clear that both benevolence and rites are important concepts in the *Analects* that are often appealed to as justification for ethical decisions, as well as promulgated as important values. But while benevolence is also accepted as an important value in the *Mozi*, the word *li* rarely appears there at all, and when it does it rarely refers to the concept of *li* itself, or to a unified scheme or collection of rites, but is generally qualified to refer to some specific type of rite being practiced at the time, such as rites between men and women, rites of marriage, and so on. The main exception to this is when it appears repeatedly in the compound "rites and music" (禮樂 *li-yue*)—the term the Mohists use to refer to elaborate Confucian ceremony, which they wholeheartedly condemn. Thus it seems that the Mohists reject not only the *content* of the rites to which the *Analects* were so attached but also the role the rites played in Confucian ethics. Though the Mohists argue for extravagant ritual mourning to be replaced with simpler ceremonies, there is no suggestion in the *Mozi* that "the rites" need to be extensively revised or replaced by new, more frugal rites; the solution is to be more radical than that —the *li* as an all-pervasive code of conduct are to be thrown out altogether.

One reason for this offered in the *Mozi* is that tradition, such as is embodied by a code like the *li*, lacks justification—the mere fact that it is tradition gives us no reason to believe that what is prescribed by it is morally right. The text gives the example of a state in which by custom "the first-born son was dismembered and devoured after birth and this was said to be propitious for his younger brothers. When the father died the mother was carried away and abandoned, it being said that one should not live with the wife of a ghost."[22] The text goes on to say that this was not the way of *ren* and *yi* but simply a case of adherence to habit and custom being taken as convenient and moral. Though *li* is not mentioned by name, the point is clearly made: adherence to a practice is not justified merely by virtue of it being tradition, even if adherents can provide superficial reasons, such as something being "propitious for younger brothers" or "one should not live with the wife of a ghost"—parodies of the kind of justification early Confucians might give for ritual practices.[23] Whatever is truly *right* must be able to stand up to some form of genuine scrutiny—and nothing is beyond scrutiny simply because it is tradition.

The Mohists effectively replace the role of the rites as a code of conduct with their own conception of normative ethics, explained in terms of not rituals but standards (*fa* 法). The Mohists explain what they mean by standards by analogy to the measuring tools of artisans: the compasses of a wheelwright and the square of a carpenter. Compasses compared to an object can be used to judge whether or not it is truly circular; a carpenter's square compared to an object tells us whether or not it is truly square. But standards do not merely apply to craftsmanship; according to the *fa-yi* chapter of the *Mozi*, "nothing is accomplished without standards."[24] Not only artisans but also generals and ministers have their standards.[25] Standards in themselves do not guarantee success in the task at hand—this is still a matter of skill—but they do improve the results of even the unskilled.

The Mohists go on to complain that those governing the world and the large states of the day do so without applying standards, thus showing themselves to be inferior to artisans. This leads them to the question of what the right standard is for ordered governance. Taking any single individual as the standard would Lead to the same problem as simply following tradition: there is no way to know that the person being followed exemplifies morally good qualities such as *ren*. Thus the Mohists commit themselves to finding moral standards that can be used in the same way as a carpenter's tools to determine unambiguously and objectively what is right and what is not.

Though rites and standards clearly have similarities, both playing a similar role in determining which actions are the right actions, Mohist standards differ importantly from Confucian rites both conceptually as well as in their content. Like the rites, standards still require interpretation. But unlike the rites, the Mohists believe that once we know what the right standards are, they can and should be applied universally without exceptions. To determine whether something is right, one needs merely to apply the standards to it, and, subject to interpretation, this will yield an unambiguous answer.

By analogy to the carpenter's tools, standards are public and (in principle at least) easily applicable by all.[26] A Mohist standard does not admit of exceptions any more than it is acceptable for a carpenter to supply his customer with a non-round wheel—if the standard doesn't fit the case at hand, everyone can, and should, declare it to be wrong. If it doesn't fit, then either the case at hand is wrong or there is something wrong with the choice of standard. Though experts in the *fa* might be better placed than ordinary people to make reliable judgments on what is in accordance with them, their judgment should be verifiable by others applying the same standard. Thus just as the unskilled wheelwright (or wheel buyer) can determine whether a particular wheel is truly round, so can an ordinary person determine whether a particular doctrine is correct; though there may be those with greater authority, an *appeal* to authority is not required to decide whether a standard fits a case or not.

THE MOHIST IDEAL OF ORDERED SOCIETY: UNIFYING THE YI

The difference between attachment to the "flexible" *li* and attachment to the inflexible *fa* has bearing on how Mohists conceive of an ordered society. Though Mohists, like many other thinkers of the day, saw finding the one correct *dao* (way, guiding path) and getting society to follow it to be their ultimate aim, their attachment to inflexible standards pushed them to be less tolerant of alternative ways. For when the standards by which society is to be governed are decided by thinkers such as Mozi and his followers, and not by adherence to centuries-old tradition, it quickly becomes apparent that the exact standards to be employed are essentially a matter of debate—indeed, much of the argumentation of the *Mozi* can be seen as defending the particular standards that the book wishes to lead society by, ultimately in terms of appeal to generally accepted values such as order and benefit to the world. The problem is that without a fixed, preexisting, and privileged standard to appeal to, the standards by which we are told we must govern the world appear not only to be in need of justification but subject to the objection that other incompatible standards might be equally well justified.

The Mohists explicitly acknowledge these problems, and their response is twofold: firstly, to attempt to justify their choice of standards by appeal to heaven (*tian* 天) as an ultimate, privileged, natural standard that can be used to justify fundamental standards such as that one should promote the benefit of all; secondly, to argue that these fundamental standards imply that we must unify people's conception of what is right. According to the Mohists, disagreement about what is right, and therefore about which are the correct standards by which to govern society, leads to disorder—and therefore such disagreement cannot be allowed. This gives rise to an authoritarian streak in Mohist philosophy: whereas Confucius might be content to criticize minor deviation from the rites as arrogant or disrespectful, and in other cases may even endorse deviation from the rites as economical or prudent, the *Mozi* must condemn all deviation from Mohist standards because allowing deviation would undermine the whole project of unifying standards and conceptions of the right and would ultimately lead to chaos. As soon as someone intentionally deviates from a standard in a consistent manner, we have two incompatible standards. A standard is something against which things can be compared to different things, yielding the same results for things that are similar to one another—it must be constant to be able to fulfill this role. If some wheelwrights use round standards for their wheels and others square ones, this will inevitably lead to trouble.

SETTING THE STANDARD

After introducing the concept of standards, the *Mozi* asks what can be taken as the standard for achieving order and gives the answer "heaven";[27] at another point we are told that following the will of heaven is the standard of righteousness.[28] This is apparently because of the vastness and constancy of heaven, because the sage-kings took heaven as their standard and also, crucially, because heaven cares for and benefits us inclusively.

But on the face of it, saying that we should "take heaven as the standard" might appear to be little better than saying that we should "do what is right" or "follow the *li* when appropriate"—in fact, it seems much *less* clear than either of these alternatives, since we do not have any sort of direct access to heaven and on initial examination it would appear that we have no idea what heaven might approve of or dislike, even supposing it is capable of approving or disliking. Worse yet, it appears to be open to vastly differing interpretations; merely saying that we should do as heaven desires might equally be used as an argument for some versions of Daoist ethics (heaven desires what is natural in the sense of "non-action" [無為 *wu-wei*]) or Confucian ethics (heaven desires what is natural in the sense of what actually is so—thus justifying existing tradition). Given the Mohist emphasis on the importance of standards, and the concrete and intuitive examples that the Mohists give of them, "heaven" seems like a very curious choice of standard upon which to base their ethical theory.

Yet although heaven clearly plays an important justificatory role in Mohist doctrine, it seems that perhaps the most *pragmatically* important standard to the Mohists is that one should "benefit the world and remove its harms" (興天下之利, 除天下之害). Though there are a number of passages that appeal to the will or intention of heaven (天意, 天志), and also a few that discuss taking heaven as the standard for order (法天), there are more passages than all of these combined that mention "benefit to the world" (天下之利). More significantly, these occurrences are very often explicit appeals to this standard as justification for why people should endorse specific Mohist doctrines such as inclusive care, establishing the existence of ghosts, rejecting music, and so on. A common phraseology is "If the rulers and officers of the day really wish to benefit the world and remove its harms, then they must..."—where what completes the sentence is the endorsement of the particular doctrine being argued for. "Benefiting the world and removing its harms" is also cited as the task of the benevolent (*ren*) person.[29] From the many examples given throughout the text, it is clear that a key way in which this is to be done is by analyzing the causes of disorder, implementing strategies to combat them, and in so doing benefiting the ordinary people.[30] Though this *fa* still requires much interpretation, it is both much clearer than a direct appeal to heaven and more frequently used in the *Mozi*; it is also this *fa* that contributes much of the content to Mohist ethics and in many ways is its most distinctive feature—it is no coincidence that many scholars see their ethical theory having similarities to utilitarianism and consequentialism.[31] For the Mohists, heaven desires righteousness and detests unrighteousness;[32] it cares for and benefits people equally;[33] and it punishes those who do wrong.[34] Thus, by taking heaven as the standard, we must do as heaven does—in other words, we must endorse the Mohist social doctrine that is ultimately grounded in heaven.

Heaven also plays a key theoretical role in explaining why, in the Mohists' proposed system of governance, society will not merely be at the mercy of a tyrannical ruler; for as a consequence of heaven's universal care for the people and dislike of unrighteousness, it will ultimately punish rulers who promote unrighteousness and disorder. Ultimately, however, it is not heaven that makes this system right but rather the fact that, according to the Mohists, this system will benefit the world.[35] When we look at what actually seems to be inspiring or determining the individual uniquely Mohist doctrines, it seems that procuring benefit and avoiding harm are central to both the motivation and justification of each. Heaven, on the other hand, seems mainly to come in to offer further justification of doctrines that have already been determined by the Mohists to be correct on the basis that they bring benefits and remove harms. Indeed, the doctrine that heaven has a positive "will" or "intention" that we should attempt to emulate at one point appears to be argued for on the grounds that accepting it will bring benefits and help us avoid harms.[36]

Why then do the Mohists assert that *heaven* is the standard all should follow? A couple of linguistic and cultural points may be relevant here. Firstly, the very common classical Chinese term used for "everything/everyone/all people" (天下 *tian xia*) literally translates as "[all] under heaven." At the same time, the word "heaven/sky" (天 *tian*) itself, which appears in this compound term, can also mean "nature/the natural world." Just as we are all people "under heaven," we can all be said to be "people *of* heaven" (天民 *tian min*). At the same time, in classical Chinese we are also said to be "born *by* heaven."[37] Therefore, Mohist doctrine aside, all people "belong" to heaven both in the sense that they are under it and in the sense that they are products of nature and thus are born by it. The wide range of meanings and connotations of the term *tian*, combined with the fact that the standards the Mohists were interested in did not have to be stated rules as such but could also be things or concepts such as *tian*,[38] led to considerable ambiguity in phrases such as "take heaven as a standard."

To "take heaven as a standard" can mean either "take heaven—the *intentions* of heaven—as a standard" or "take heaven—in the sense of inclusivity of all, everywhere, everyone—as a standard"; in other words "take *all* as a standard." In one chapter of the *Mozi* we see precisely this ambiguous usage in the context of a discussion that primarily seems to hinge on benefit and otherwise uses the word *tian* only in the compound *tian xia*: "Is this intended to benefit heaven? But this is to take the people of heaven and with them attack the towns of heaven; this is to murder the men of heaven.... It is then not a benefit to heaven."[39] Heaven seems to be invoked primarily to emphasize that the scope of what is at stake is "all under heaven"; the argument being presented does not otherwise hinge upon any particular peculiarly Mohist assumptions about heaven—only the fact that we all in some sense belong to it. In the same sense in which "heaven's people" and "heaven's towns" refer to all people and all towns, so "benefiting heaven" is essentially another way of saying "benefiting all." Thus when the Mohists say, "heaven is inclusive," it is worth remembering that there is a linguistic background to this claim. Heaven *is* inclusive—in the sense that "heaven's people," "heaven's towns," and so forth refer to *all* people and *all* towns.

The Mohists emphasize and develop this point: we are all equal before heaven—not in any particularly religious sense but simply because in established senses of the word, all of us are born by it, living under it, and subject to it. If the points the Mohists do argue for regarding heaven's inclusiveness seem strange—that it loves and benefits everyone equally, for example—it is worth remembering that there were many aspects of heaven's inclusivity that would have been generally accepted by their contemporaries and so would not be seen to require further justification.

The key Mohist invention with respect to heaven seems to be viewing it as the culmination of a natural hierarchical progression from ordinary people, through various levels of leadership, up to the son of heaven, and finally to heaven itself. In the Mohist ideal society, in this hierarchy the leader at each level is to be the most benevolent person from the level below: the village leader is to be the most benevolent person of the village; the provincial leader the most benevolent person of the province, and so on.

What justifies this system is that each person is the most benevolent from a pool of those below him in the hierarchy, and thus each superior will be at least as benevolent as those beneath him. This gives each leader the moral authority to rule over and unify the *shi-fei* (是非 this/not this or right/not right) attitudes of those below him in accordance with the wishes of those above, at least in part using rewards and punishment. But the symmetry of this system leads us up to the son of heaven, who, though his position is justified by his supposedly being the most benevolent person in the world, apparently has no one remaining to appeal to for guidance or to reward or punish him should this be necessary. Given that *tian* (in the sense of "nature") is already the source of unforeseeable and uncontrollable disasters and blessings, it is only a short step to see that heaven would make an excellent candidate for his superior—he is in any case already "his son" (天子 *tian zi*). Having gone this far, in some sense the symmetry of the system now requires us to ascribe benevolence to "heaven"—and this is exactly what the Mohists go on to do. Combine the various connotations of the word *heaven* in classical Chinese discourse with its role in the idealized political hierarchy and with the Mohist aim to bring benefits to the world and remove its harms, and we already are committed to something very similar to the Mohist conception of heaven and its role as an ultimate standard.

APPLYING THE STANDARD

The Mohists frequently come across as being genuinely exasperated that people of the time, despite being made aware of what the Mohists saw as highly persuasive arguments, failed to accept and follow Mohist doctrine.[40] Whereas a cultured appreciation of the content, background, and deeper meaning of the rites would be required for a person to be a *junzi* (君子 gentleman, superior man) or *shengren* (聖人 sage) for the Confucians, for the Mohists all that is required to be a model member of society is that one accord to a set of fairly straightforward standards. Whereas Confucians seemed to implicitly accept that not everyone would have the ability to follow their *dao*—not everyone would become a *junzi*—the Mohists saw everyone as capable in principle of following their *dao* and believed that the reason they did not was simply that they had not been given adequate moral education. The Mohists most often described the failure of their contemporaries to follow their proposed *dao* not in terms of refusal, disagreement, or lack of interest in doing so but rather in terms of failure to know right from wrong.[41]

Without the emphasis on classic texts, music, ritual, and poetry advocated by the Confucians, Mohist doctrine would have been much more intelligible to the masses than the Confucian rites, which appear in any case to have been aimed primarily at the educated elite. The *Analects* records that Confucius was often dissatisfied with the words and actions of many of his disciples, and though his disciples were presumably among those most sympathetic to his teachings and most eager to follow them, it seems apparent that the majority of them failed to do so sufficiently well—sometimes due to lack of motivation but also due to lack of ability or intelligence.[42] Not only their actions but their understanding of Confucius's moral system sometimes seem to be called into question.[43] The Mohist maxim—repeatedly appearing in identical wording throughout the *Mozi*—that one should in one's actions always try to "bring benefit to the world and remove harms from the world," though of course subject to interpretation, contrasts starkly with the widely varying guidance Confucius offers to disciples when they inquire about benevolence, despite this being perhaps the most important value discussed in the *Analects*.

STANDARDS, UNIFORMITY, AND ORDER

The Mohist focus on standards brings with it an interest in uniformity and consistency. A key feature of much of the moral reasoning that appears in the *Mozi* is its rejection of exceptions: if someone deems something right in almost all cases and then wrong in another relevantly similar case, the Mohists see him as being inconsistent in his reasoning and actions. This brings us back to the Mohist idea of a standard as something that can be used to determine unambiguously what is and what is not: standards admit of no exceptions. Even allowing that others might have different standards, the Mohists believe that the only correct *way* to make moral judgments is to use a standard, and therefore the same standard, whatever it is, must be applied in each case. Those who fail to see this are accused not merely of being wrong but of "knowing the small, yet not knowing the great"[44] and "on seeing a little black, calling it 'black,' yet on seeing much black, calling it 'white'"[45] —failing to apply their own standards, whatever they might be, in a consistent manner.

With respect to their own standards, although the Mohists were not interested in reforming basic social structures, and in fact affirm at length the existing hierarchy of society, they do attempt to bring a certain inclusiveness and consistency to their code of conduct in many of their key principles. For example, inclusive care applies to everyone and in principle requires care for everyone; rejection of aggression is argued for precisely on the grounds that it is inconsistent to say that a man murdering his neighbor and stealing his property is wrong and yet not to say that aggressive warfare is also wrong; wasteful expenditure is wrong for a town leader just as it is for the ruler of a state.

What standards bring to the Mohist moral code is not so much equality in the sense of inherent equality of people as individuals—people have unequal social roles, and the son of heaven certainly has more power than an ordinary man—but equality before a common moral code that must be consistently applied to all, regardless of their role in society, without exception. The same standards apply to all, without exception, even for those, such as the son of heaven, who might traditionally have been seen as exceptional cases.

Again this contrasts with Confucian thinking, where there appears to be an acknowledgement that the rules governing the educated and ruling classes can be quite different than those for the masses. Certainly, the rites seem to be primarily intended for those higher up the social hierarchy as opposed to the common people.[46] Talk in the *Analects* of how "lesser men" behave can perhaps be dismissed as talk of morally inferior people who would not exist in an ideal society because they would all become *junzi*. It might be argued that in texts of the period the term *junzi* had acquired a primarily moral meaning and no longer carried class overtones—though on such a reading, when the *Mengzi* proclaims that "the *junzi* should stay clear of the kitchen,"[47] one wonders who it supposes should be left to cook. But talk of the "ordinary people" (民 *min*) in the *Analects* seems less easily dismissed. One passage states that ordinary people can be made to follow but not understand;[48] many passages talk of *making* (使 *shi*) the common people do various things, the just doing of which is apparently one aspect of the *junzi*'s *dao*.[49] It seems at least that there is some expectation that different classes of people will need to be assessed by different standards: what is expected and required of the *junzi* will not be expected and required of the common people.

For a Mohist, however, the standards remain the same no matter who one is: it is *never* right for anyone to deviate from what brings the greatest benefit to the world. This devotion to absolute standards not only influenced the policies the Mohists advocated and the style of their argumentation but also may have provided the theoretical support for the fanatical devotion to the Mohist social project testified to in many contemporary historical sources. As the Mohists say, people will fight to the death over a matter of principle: nothing is more important than righteousness.

CONCLUSION

The Mohists started out with two key ideas: promoting benefit in the world, and finding objective standards that would enable people to agree upon which doctrines would actually do so. These desires may well have been motivated by the political and social realities of the Warring States period and dissatisfaction with the social and moral codes that to some degree underwrote these realities. They advocated a conception of order that replaced the perceived malleability of Confucian doctrine with a fixed moral code by using rigid standards in place of somewhat malleable rites.

In doing so, they opened up a debate over what the correct standards are and what justifies them.

Seeing the dangers of conflict over choosing the right standards, they appealed to a rigid structure of governance in which the single, correct standards are set by one's immediate superior, leaving the ultimate ruler, the son of heaven, as the final human arbiter of standards. Recognizing that rulers are not always as altruistic or good at promoting social order as one might hope, they saw heaven as the source of an ultimate guiding standard with which to complete the system. This arrangement was not chosen by chance or merely out of some prior religious devotion to heaven. Rather, given the existing Chinese conception of heaven, the Mohists' affirmation of the existing social hierarchy, and a commitment to finding universally applicable and objective standards that would benefit the entire world, heaven was an obvious choice for the *ultimate* standard by which the world should be ordered.

In an idealized Mohist or Confucian society, everyone is benevolent in his or her actions. But whereas in a perfectly ordered Confucian society, most people follow the *li* most of the time, in a perfectly ordered Mohist society, everyone follows the correct *fa* all of the time. The Mohists succeeded in developing a more consistent ethical code and in highlighting the hypocrisy of others of their time, but this came at the price of instituting an authoritarian conception of governance in which whether or not righteousness truly prevailed might ultimately be decided by heaven.

CHAPTER 4

Hegel, Lao Tzu, and Bohr: The Merging of Traditions

Robert Elliott Allinson

Recently, through an important letter, it has been demonstrated that Niels Bohr, the father of quantum physics, was acquainted at a young age with the ancient Chinese philosophy of Lao Tzu. New evidence has come to light from a letter written by Bohr himself, discovered in 1988 by Finn Aaserud, a noted historian of science and director of the Niels Bohr Archives. The letter is a reply to a letter of inquiry from a Svend Hugo Jurgensen, a teacher from the Danish town of Aalborg. The teacher had sent Bohr a manuscript entitled *Tao Te Ching and the Idea of Complementarity*. Bohr's reply, dated March 26, 1958, begins, "I thank you for your letter and the enclosed little note about Tao Te Ching, which I have read with great interest. I believed what you say about the old Chinese philosophy is in many ways quite to the point. *In my youth I received a beautiful impression of it through Ernst Miller's book 'Oldmeister,'* and during a visit to China twenty years ago I learned how highly the memory of Lao-Tzu is still valued." One must bear in mind that "Oldmeister" is a Danish rendition of the name Lao Tzu and that the book itself is a translation of and commentary upon the manuscript left by Lao Tzu. This may well have been the very book Bohr's son Hans was referring to when, on October 17, 1995, at the Niels Bohr Institute, he told me that his father had been acquainted with Lao Tzu in a Danish translation.

There has been discussion as to whether Eastern philosophy influenced Bohr's discovery of complementarity in physics. The idea of the complementarity of opposites is advanced in the *Tao Te Ching*. (It is of course also an integral part of the teachings of the *I-Ching*.) Bohr's acquaintance with Lao Tzu is both early and authentic. I am indebted to Finn Aaserud both for bringing Bohr's letter to my attention and for providing the translation from the Danish, above. The original letter can be found in the Niels Bohr General Correspondence file, housed in the Niels Bohr Archives in Copenhagen. Since Miller's book was first published in 1909 (the year Bohr turned twenty-four) and since Bohr by his own account was in his youth when he read it, it is clear that his knowledge of Chinese philosophy preceded his discovery of the complementarity principle in physics (in 1927).[1] In fact, when Bohr first announced his complementarity principle, scientific colleagues who had studied with him said that it was the same thing he was talking about while studying in university.

While all of this evidence does not conclusively prove that Bohr learned his idea of complementarity from Chinese philosophy, this letter, which includes the intriguing sentence "The old Chinese philosophy is in many ways quite to the point," makes us wonder what many ways in Chinese philosophy were "to the point." The discovery of this letter of Bohr's is a discovery of great value.

Would it be only a strange coincidence that Bohr's brilliant complementarity principle, which had such importance and influence for Western physics, bore such close resemblance to the ideas of Chinese philosophy that Bohr had imbibed from his youth? And in fact, even if, despite the evidence, Bohr's discovery of the complementarity principle owed nothing at all to his reading of Lao Tzu in his youth and the similarity was pure coincidence, the fact is that we still have a wonderful illustration that the idea of complementarity is universal in its intelligibility and that the Chinese mind is not of a different order of humanity, since it either was accessible through the medium of the Danish language to a Western physicist or the Danish physicist was able to understand and apply a concept that was originally discovered by a Chinese mind. That the same Western physicist could then apply the idea of complementarity to understand the workings of the universe common to both Western and Chinese minds is a demonstration that what is understood is something universal and is solely Chinese only insofar as empirically it was first understood in Chinese civilization, in the same way that the invention of printing occurred in China but possesses universal applicability. To put it another way, the Chinese mind is as universal in its application to reality as its Western counterpart. What better evidence can be cited to show that the Chinese mind is intelligible to a Westerner than that a Western physicist was able to draw from its obscure and subtle principles major lessons for Western science?

Bohr's fascination and identification with Chinese philosophy was profound. On Hanna Rosental's advice, when Niels Bohr was knighted, he chose the T'ai-chi symbol for the emblem of his coat of arms with the accompanying motto "Contraria sunt complementa" (Opposites are complementary).[2] Is it not remarkable that one of the greatest Western physicists of the twentieth century was so influenced by classical Chinese philosophy that he chose for his emblem and for his motto for his coat of arms, by which he most surely would be forever remembered, a classical Chinese image and a classical Chinese leitmotif, both of which represent a core perspective and a core value of ancient Chinese philosophy? In addition, Bohr's favorite quotation from poetry was from Schiller's *Sayings of Confucius*: "Only wholeness leads to clarity." How highly did the Western physicist, whose discoveries in quantum physics did so much to change the face of Western physics in the twentieth century, value ancient Chinese philosophy? Not only was the discipline of Chinese philosophy intelligible to the great physicist Niels Bohr, who was in a different discipline altogether, but he was able to apply the insights of this different discipline to break new ground in the archetype of Western thinking, the science of physics.

There are two interdisciplinary leaps here: from culture to culture and from humanities to science. Even if one wishes to argue that Bohr's discovery of complementarity was coincidental and not inspired or influenced by Chinese philosophy, what is established beyond question is that interdisciplinary understanding is both possible and can be extraordinarily fruitful. Finn Aaserud, in fact, related to me that Bohr was more of a philosopher than a physicist.

Indeed, as Christian Bohr, one of Bohr's grandsons, related to me, the dragon, symbolizing wisdom, was originally proposed as the symbol for Bohr's coat of arms when Bohr received the Order of the Elephant. But this was turned down on the grounds that the dragon was not a heraldic animal. As a counteroffer, it was proposed that stars be placed inside the four corners of the figure on the shield on the center of which the T'ai-chi symbol was to be placed, but Bohr turned down this idea on the grounds that stars could not be inside the cosmos. Thus, unlike all the other coats of arms in Fredericksberg Castle, in Bohr's native Denmark, the emblem on Bohr's shield has nothing in each corner of the figure in which the T'ai-chi symbol is placed. Inside the universe it may well be supposed that Eastern and Western views of philosophy somehow could also coexist.

How interesting it might prove to be if Bohr's metaphor for explaining the differences that exist in the microscopic world of physics might also prove invaluable as a metaphor for the explanation of the contrasting mental sets that make up the intellectual universe. Just as Bohr did not intend his explanation to be a description of ontology (as, it can be argued, Lao Tzu did before him), it is not necessary to intend that an explanatory model for Eastern and Western modes of thinking is an ontological description.[3] If a physicist may be allowed a metaphor for the explanation of the physical universe, then a fortiori a philosopher may be allowed a metaphor for a description of the mental universe or the world of explanations.

What is of course obvious from Bohr's choice of a metaphor is that there is good evidence to demonstrate that this metaphor most likely arises from or at the very least is compatible with the soft, ruminative discipline of Chinese philosophy, and yet it is a core principle for quantum physics. His metaphor of complementarity is the expression of the basic principle of the yin-yang philosophy. If a metaphor taken from Chinese philosophy may be of great explanatory power in the world of physics, may it not be all the more possible that the same metaphor may possess great explanatory power in the world of philosophy?

This might represent a case in which a regulative principle of physics can be utilized as a regulative principle of philosophy. In the sense that Kant called his breakthrough ideas a Copernican revolution in philosophy, we may consider the application of the complementarity principle to philosophy a Bohrian revolution in philosophy. It is of special interest to note that Bohr himself held that the greatest significance of the ideas of physics for philosophy lay precisely in the implications that new ideas of physics possessed in challenging the foundations of our most fundamental concepts.[4] A concept that has once left home, mimicking in physics the exact fate of the Danish philosophy of Kierkegaard, after it has been celebrated abroad may be accorded the respect that it did not receive while it stayed within its own disciplinary boundaries.

For the present discussion, the key dimension of the complementarity principle in physics is that the two viewpoints, whether matter is perceived of as wave or particle, are considered to be harmonious viewpoints. In other words, Bohr's explanatory model was that the core structure of the universe is not perceived of as self-contradictory but as harmonious, even though composed of different and contrasting ingredients. Such a model provided Bohr with the greatest possible explanatory power in the world of atomic physics. The *two different and contrasting, although not competing*, models of wave and particle *completed not competed* with each other to form a complete union or whole. That each model did not compete with but rather complemented the other in the composition of the whole is what led Bohr to the famous name the complementarity principle.[5] In certain circumstances, one model was to be preferred as the explanatory model; in other circumstances, the other model was to be preferred. But *neither model conflicted with the other, and neither model struggled with the other for ultimate or sole supremacy.* Interesting enough, one model was collective (the wave); one particulate (the particle). Even the very composition of the two models mirrored the composition of the models of the self respectively of East and West. (It is not to be suggested that the content of the models was self-consciously adapted from the images of the self of East and West by Bohr, but it is not a surprising discovery.)

What is most amazing is that while complementarity, a philosophical notion, has been embraced by physics to harmonize differing accounts of the physical universe, back home, *complementarity has yet to be embraced by the philosophical community to harmonize different accounts of the mental or philosophical universe*. How far philosophy lags behind physics! It is my hope that this paper may provide a stimulus for the application of the complementarity principle to its own discipline.

The notion of harmony as the underlying basis for coexistence is an ancient and arguably the most central notion of Chinese philosophy. Wing-tsit Chan, the late distinguished Chinese philosopher in the West, stated,

> The foundation of the Confucian system lies in the moral realm that is in human experience itself. The thread is also generally taken to be identical with the Confucian doctrine of central harmony (*chung-yung*, Golden Mean). Indeed, this doctrine is of supreme importance in Chinese philosophy; it is not only the backbone of Confucianism, both ancient and modern, but also of Chinese philosophy as a whole. Confucius said that "to be central (*chung*) [with all]" is the supreme attainment in our moral life. This seems to suggest that Confucius had as the bias of his ethics something psychological or metaphysical.[6]

When the achievement of harmony is explained as a result of the understanding that differences are complementary rather than conflictual, it becomes clear why a model that posits conflict and competition at the basis of reality is most likely to lead to disharmony and chaos as a result. The complementarity principle may be taken as the ultimate model, not only for physical explanation and for the achievement of harmony in Chinese philosophy—nay, in life—but also as the guiding principle for understanding East and West mentalities. A Chinese philosophical principle can be the central principle of explanation in physics, in life, and in the understanding of contrasting and diverging points of view.

To further unfold the concept of complementarity in Chinese philosophy, one may understand the concept as referring to a means of understanding change. An illuminating way to understand the Chinese model for understanding change is to contrast it with arguably the most powerful, influential, comprehensive, and effective Western model of understanding change: the Hegelian dialectic.[7] For Chinese philosophy, the two halves, yin and yang, which are perceived to be two halves that make up the whole, are in fact related to each other in part by a cyclical process of change. In the Hegelian dialectic, in the process of *aufhebung*, the new concept replaces the old concept as the old concept is negated, although some of it is preserved in the new concept. The image of the Hegelian progression is of a spiral moving ever upward as new concepts replace old ones while including parts of the old ones within themselves.

In the yin-yang progression, the two concepts exist simultaneously with each other, and while one gradually replaces the other, the replacement is only temporary: the one that has been replaced gradually regains its ascendancy. The image of the progression is a circle in which the top and bottom halves rotate in terms of their ascendancy and descendancy, but each half is never entirely replaced by the other half. The progression therefore is the rotation of a circle and not the ever-upward movement of the spiral.

A further difference is that in the Hegelian dialectic, the two concepts are in warlike opposition to each other such that there is an antagonism between them. In the yin-yang progression, the two concepts are not antagonistically opposed but are both necessary to one another's existence and complement and to a certain degree constitute one another's existence. In the Hegelian dialectic, there is an infinite succession of new concepts replacing old and inadequate ones (leaving open the obligatory question as to whether Hegel's own system achieved the final progression and hence in a way negated the concept of the infinite process); in the yin-yang progression, there is a constant rotation between two sides of the same concept, revealing the necessity of both halves to form a greater whole, which at certain times emphasizes one of its aspects and at other times emphasizes the other.

The yin-yang progression follows a phase of expansion and contraction, like the phases of the moon, in which one side of the concept reaches its fullness and thus reaches its fruition and then must descend to allow the other side of the concept to dominate for a time. In the Hegelian dialectic, there is a constant onward progression of new concepts and there is no corresponding notion of phases of ascendancy and descendancy.

In the yin-yang progression, the two sides are not replaced by a third, but each side requires the other side for its own completeness. The two sides gradually replace and are replaced by each other in terms of ascendancy and descendancy; there is a phase or period during which it is correct that one be at the zenith and also a natural time for it to recede to the nadir and to be replaced by its other half.

After this explanation, the notion of how to apply the principle of complementarity to East/West "comparative philosophy," or what is better termed integrative philosophy, as will become clearer below, becomes more apparent. Firstly, one acknowledges that Eastern and Western philosophical approaches, or mental sets, are both bona fide dimensions of the human mind and that each approach can be called upon whenever its unique merits best address or solve problems that arise relevant to the human condition. For this dual application theory to operate, of course, it must be granted that each side is transparent to the other. If the same human being is to possess the skill to choose the approach that best fits a co-temporary condition, then both sides must be equally available to the same human being. In a way, then, the label "comparative philosophy" is a creaky old description that needs to be abandoned, summoning up as it does a picture of two static, ahistorical images existing side by side, left to stand alone to be passively compared and contrasted with each other.

Instead, the labels "integrative philosophy" and "complementary philosophy" are to be preferred, since the object will be to select whichever dimension of philosophy best addresses problems that are arising. The label "integrative" emphasizes that both yin (Chinese philosophy) and yang (Western philosophy) make up a more complete union; the label "complementary" emphasizes that East and West coexist, add to each other, and await their selection in harmony as suits the needs of the time. The notion of "integration" possesses the advantage of stressing the constant and continuous need to bring both philosophies into play and into harmony with each other; the notion of "complementarity" possesses the advantage of stressing that either philosophy may be chosen by the world philosopher as the philosophy of choice for the co-temporary moment. Because of the unique advantages of each label, it is most advantageous, and most suitable in the spirit of complementarity, to adopt the custom of utilizing both.

Needless to say, such a slight description as appears above is an oversimplified account. The notion that philosophy can be encapsulated by such a geographical metaphor is misleading and certainly not all inclusive. However, given the limitations of such categorization, such a classification possesses at least a pedagogical usefulness. In reality, a philosophy or mental set may consist of an intricate combination of both Eastern and Western (and other) emphases in a subtle and delicate balance. Yin and yang are never in complete separation, and the distinction of relative ascendency or descendancy may be a matter of the most delicate of degrees. At times one may dominate completely; at other times, the other dominates with equal force and intensity. At still other times the blend is incomparably woven in such a tight knit that it is nearly impossible to tell which is which. At yet other times the mixture breaks apart, and one tendency gradually intensifies in a crescendo while the other gradually diminishes in a decrescendo. At even other times there will be such a dizzying rate of change, in which one tendency replaces the other in rapid succession, that one can experience only the most staccato of rhythms. And at other times one can experience even atonal combinations of the two, which, as in a Stravinsky composition, can issue forth in the most mystical and luminous of East/West harmonies.

The model of "comparative philosophy" suggests a static witness role for the philosopher, who stands on the sidelines and comments, as a neutral journalist or television newscaster, on the merits and demerits of Eastern and Western mentalities. Under this static model, the concept of understanding is reduced to the concept of tolerance. But tolerance is not really understanding; tolerance is a tolerance of differences that are left mutually abiding and possibly troubling. In fact, the concept of tolerance can even include a touch of arrogance and perhaps condescension, as in "we can tolerate those people." The concept of tolerance also entails a passive model of cross-cultural philosophic no growth, since the connotation of tolerance is that the tolerant one already possesses a fully formed viewpoint that is the "right" viewpoint—and all the more "right" since it tolerates other viewpoints different from itself. The concept of tolerance is suspect because it carries with it the hidden prejudice that the tolerant party implicitly possesses the right point of view.

The viewpoint of "scientific objectivity" is also one of which one should be appropriately wary. The notion of the "scientific observer" carries with it the prejudice that the view that abstracts from emotions is more valid than the one that is emotion laden. Another problem is that such a view precludes an alteration of one's own viewpoint and hence prevents one from enriching oneself from the tolerated viewpoint or from going in another direction altogether.

It should be stressed that it is not altogether fair to describe the scientific viewpoint as emotionally neutral—that is, devoid of emotionality—as this is not accurate, strictly speaking. The scientific viewpoint is not emotionally neutral (it is difficult to know what this means) but is rather emotionally "cold" or emotionally detached. Emotional coldness or detachment is a certain kind of emotionality, one that the British philosopher, Collingwood satirizes in his *Principles of Art*.

More than two decades ago, in the introductory chapter to my book *Understanding the Chinese Mind: The Philosophical Roots*, the model of a proactive immersion into each culture was proposed as a viable hermeneutic for understanding that East and West are not alien to each other, as each represents a different emphasis and degree of development of a tendency of the human mind.[8] The why of such a difference was also the subject of that inquiry and cannot here be recapitulated. Suffice it to say that it was to a certain extent historical contingency that the East and the West developed in different directions such that each represented a complementary hemisphere, as it were, of the human mind.

If one divided the globe into eastern and western hemispheres, the globe could be construed as the macrocosm of the human brain, the microcosm of which Chinese culture sees in the walnut. The eastern half (the right side) can be held to comprise the holistic, imaginative, intuitive side; the western side can be held to be the dichotomizing, distinction making, logical side. For a fully functioning human brain or mind, both sides must be integrated so that one can operate at peak or optimal efficiency and actuality. Indeed, the dexterity and deftness with which one can go from one side to the other is a measure of how developed one can become as a human being.

The problem of a view that does not benefit from the assistance of a complementarity principle or a principle of harmonization is that one is forced into an either/or kind of thinking. Either one view is right or the other view is right; there is no space for both views being correct. With either/or thinking, one is forced into combat when one comes into contact with a view that is different, for all differences must be perceived of as conflicting differences.

It is suggested herein that the model of integrative or complementary philosophy is that one is not limited to the role in which history has cast oneself. One can expand one's own viewpoint; indeed, one can jettison nonserving and nondeveloping viewpoints. In Hegel's model, dialectical change involved alteration, cancellation, preservation, or sublation and creation or synthesis. While this model was conceived of as an explanation of change, it can also be perceived of as a metaphor for intellectual comparisons and contrasts—as a model, in short, for the history of philosophy. After all, this was its intellectual origin. Aristotle had already employed a "Hegelian" dialectic in his treatment of earlier philosophers. He argued in his *Metaphysics* that the pre-Socratics were attempting to say what he was saying but because of their limitations could say only lispingly.[9] Was this not the origin of Hegel's concept of the dialectic of the history of philosophy (despite Hegel's own attribution of the origin of his notion of dialectic to Plato's *Parmenides*)?

Comparative philosophy today, if it does not at least reach its Hegelian potential, remains a limited and inessential tool, a tool that is of import only to philosophical antiquarians who relish the thought of comparing and contrasting viewpoints for the pleasure of viewing them in their pristine display cases side by side, as they are observed untouched, in the intellectual museum in which they are housed.

But antiquarian philosophers also remain untouched by the viewpoints they have safely ensconced in their locked showcases. What assistance is such an idle viewing activity to either the philosopher, the case of philosophy, or the world? What assistance is the talk occasioned by such viewing? It can be likened only to intellectual gossip or, in its use of technical terms and foreign words, to a philosophy of those who talk about philosophy in the same ways as those who attend musical performances to "see and be seen."

The model of integrative philosophy, which differs from the Hegelian dialectic in the ways suggested above, suggests a proactive role for the philosopher who commits herself or himself to the adoption and life choice of a philosophy that incorporates both Eastern and Western dimensions. Understanding is based not on "tolerance" but on the experience of a way of thinking and a lifestyle that is steeped in the best of both Eastern and Western ways. To truly integrate an intellect, just as to truly integrate a society composed of different races and cultures, one must not leave the viewpoints in segregated "intelligible" and "unintelligible" categories. Not only does this provide the fuel for a later conflagration, but it is patronizing and self-limiting. To divide cultures into two alien and mutually impenetrable spheres abstracts from the model of human growth and replaces it with branches that attempt to grow on their own without thought to the human tree to which they belong.

If one adopts the view that one should wait on the sidelines and teach and practice comparative philosophy from some presumably neutral standpoint, it will be difficult if not impossible to resolve the differences that it will inevitably arise. The problems that face humankind if such differences wax into conflicts and then into wholesale atrocities are immense and morally staggering. Far better for the world if Eastern and Western philosophy were to be integrated into a more complete whole, a complementary field in which genuine interaction rather than polarization were to be the focus of the philosopher-intellectual.

Consider the viewpoint of such a figure as Huntington, who argues that cultures are headed for inevitable collisions.[10] If one's view of the different cultures of the world is such that they are perceived to be in inevitable conflict, then how much sincere and persistent attention will be given to attempting to provide the conditions for a life in coexistent harmony? Probably not very much. In greatest likelihood, more attention will be given to preparation for disharmony—that is, preparation for war.

Without a complementarity principle as a guiding principle of integrative philosophy, one is constantly forced into the conformational posture of either/or. Chinese philosophy can play a marvelous role for integrative philosophy in the world today by offering a principle of integration, the complementarity principle. Perhaps this is one of the leading roles Chinese philosophy can play in the intellectual world today.

The integrative, rotation model, which differs from the Hegelian dialectical model, simply suggests that the particular richness and unique contributions of both East and West must be allowed to take their places in turn to address the peculiar maladies of the historical epoch in which one finds oneself on earth. The complementarity model suggests that neither East nor West possesses a privileged standpoint but that both viewpoints are necessary to balance and complete each other to form a more perfect union. The alternative is a conflict model, and this is the way of Huntington and the inevitable collision and clash of cultures. The concept of harmony that so enriches and informs the Chinese tradition is the only viable wave of the future.

In the dialectical collision, the resolution is achieved by three stages of conceptual transformation, a process that more or less occurs simultaneously, although it is described as if it occurred seriatim. To reiterate, the first stage of the process is to jettison what is no longer relevant within the two opposing views. The second stage of the process is to salvage what is of value. The final stage is to synthesize what is remaining of the two viewpoints to form a third viewpoint that contains something of the two previous viewpoints and yet also forms a new viewpoint of its own that transcends either of the previous viewpoints. The emphasis in the Hegelian model is confrontation, destruction, preservation, synthesis, and the creation of something new. One is well reminded of the roles of the three Hindu deities: Siva, the destroyer, Vishnu, the preserver, and Brahma, the creator.

The emphasis in the Hegelian interactionism of viewpoints is on the transformation of each participating viewpoint with a view to creating a third standpoint that will resolve the previously conflicting viewpoints. Thus the goal of the interaction is to create a shared standpoint for both interacting standpoints, which is better than the standpoint that either severally possessed. This goal of achieving progress, or a better viewpoint, reveals a difference from the *I-Ching* standpoint *simpliciter*, which more closely describes a natural process.

In the model of yin-yang interactionism, the two viewpoints replace one another in turn. This replacement is only temporary, because the temporal model of the process is cyclical. Nevertheless, it is not a complete replacement because the replaced viewpoint is always necessary and in fact is a constituting element of the replacing viewpoint. The replacing viewpoint always retains a portion of the replaced viewpoint so that the replacement is always a matter of degree. When the replacing viewpoint is, in turn, replaced by the previously replaced viewpoint, it too lingers behind in a matter of degree. No viewpoint totally replaces any other viewpoint, so the mixture is always a mixture of two viewpoints. The difference is one of only degree and not one of kind. In addition, the two viewpoints are not perceived as in conflict with each other but as being in collaboration with each other.

Thus the total viewpoint that one embraces at any moment is always a combination of both viewpoints. The only difference between one moment and another is that the combination will contain both viewpoints in different degrees of ascent and decline. Differing viewpoints are perceived as needing each other in order to constitute a complete whole. The two viewpoints collaborate, as it were, to produce the relevant standpoint that is perceived as most suitable to a current situation. There is no idea of simple linear or vertical progress. The model of change is that of a circle in which the merging of viewpoints is the obtaining of a complete harmony of differing viewpoints.

While self-alteration may be considered an overarching motivation, the self that is altered is a self that expands and contracts, so to speak, rather than a self that changes altogether. Its opposing sides are not so much in conflict, demanding a solution that is different than either one alone, but rather are colleagues that rule in turn, leaving the other as a shadow cabinet until such time as its services are desired. The model is thus one of collegial cooperation rather than self-alteration.

How then to reconcile the two differing integrating modalities described so far? Should one utilize a Hegelian model of self-alteration or a yin-yang model of collegial interactionism to sort out how to structure the process of interaction? This question is indeed difficult to answer. Perhaps a key to answering this question is to consider more closely two aspects of the yin-yang model of interactionism that have not yet been fully explored. One aspect is that the qualities of yin and yang are modeled after the feminine and the masculine principle. In the Hegelian dialectic, there is no such gender parallel. When yin and yang are in collaboration, there is the driving force of the attraction between opposites.

This masculine and feminine aspect of yin and yang may perhaps offer a clue as to how to collate the Hegelian with the yin-yang models of interaction, which depends to a large extent upon the needs and conditions of the time. If the needs of the time are such that a greater harmony between differing viewpoints is desirable, the yin-yang model of interactionism may be preferable over the Hegelian model of seeking self-alteration.

Alternatively, if the needs of the time are such that a greater diversity of viewpoints is desirable, the Hegelian model of seeking self-alteration and new standpoints might be more desirable. However, as such new standpoints are adopted, they will become richer and richer in the sense that they contain more and more diverse elements retained from previous standpoints, and as a result the new standpoint achieved might, in turn, require a further reliance upon a yin-yang model of harmonizing its parts to achieve greater stability. This in turn may break down and result in the need to reach out to find new viewpoints once again. In truth, the yin-yang model of interactionism is no more static than the Hegelian model.

The third model, which contains both the Hegelian model of new standpoints and the yin-yang model of forces in harmony with each other, may function as a philosophical ombudsman. This may appear to be more a Hegelian than a yin-yang resolution. However, this is not completely the case, as while the new standpoint emphasis might be a Hegelian element, the constant need to harmonize the new viewpoint with its own incorporated parts would be a yin-yang element.

There is another element of the yin-yang viewpoint that might prove very fruitful for the future. In the yin-yang perspective, the yin and the yang are complementary to each other. Whatever process is more emphasized at any point in history—the reconciliation of the parts with each other or the reaching out for the forging of a new viewpoint—both of these aspects must also be seen as complementary to each other for a complete picture to be formed.

In this sense, it could also be said that the answer to the question as to which model to utilize, yin-yang or Hegelian, is both. One never really chooses one or the other; one always chooses both in a matter of degree. The question as to which to choose already assumes a Hegelian prejudice, as if an absolute choice is to be made. The answer of both implies a yin-yang model of harmony rather than a Hegelian model of change. In this sense, it appears preference is given to the yin-yang model.

However, the element borrowed from the Hegelian model is the seeking of a third perspective in which the differing viewpoints can join and thus reconcile their differences. In a sense, the model of progress is incorporated in this integration of yin-yang movement *simpliciter*. It may be said that an entirely new model constituted by both approaches has been constructed. This model may be symbolized by a spiral that sometimes returns upon itself, sometimes moves upward, and sometimes moves downward. While this may appear to be more of Hegelian than a yin-yang resolution, it must be stressed that it is not always an upward movement but sometimes a downward movement that is required. Thus progress is reconciled in some sense with harmony.

It is difficult to offer a pictorial image, as sometimes one must reach back to the past to discover inspiration for the future. The very question as to which aspect of the model is dominant is answered differently depending upon the needs of the time. It is this third aspect that does not lend itself to pictorial representation. One may think of the caduceus, the two snakes intertwined with each other that are the Apollonian and traditional symbol of the medical arts. In this sense, the two methods, that of East and West, are forever bound to each other. It is of interest that such a bond makes use of the symbol of the medical sorority or fraternity, which implies that the ultimate goal of philosophy is to be or possess a kinship to one of the healing arts. Was this, in the end, the meaning of the famous question of Western philosophy: Why were the last words of

Socrates "Pay a cock to Asclepius," the god of physicians? In this case, the answer reveals that Socrates's last thoughts on the subject of philosophy were that it owed a debt to the science or art of healing. This final utterance of Socrates reveals that he viewed philosophy as owing a debt to the medical arts, and this symbolic payment was a way of showing that philosophy was paying homage to the ancestor of its own divine right and calling. Without the art and science of healing, philosophy would have come into being as the physician, as it were, of the soul.

It would be appropriate, after a reference to such an exalted calling and self-induced obligation, to enter into a discussion that justifies such high-standing references. It would be of special interest to take note of a little-emphasized feature of the Hegelian dialectical model of transformation, which is the transcendence of the empirical dimensions of history and the better-known focus on the development of a new, integrated standpoint that transcends the limitations of the previous correlated standpoints.

The fundamental feature of the yin-yang developmental model that one may draw upon is the background awareness that whatever standpoint is chosen as the most needed co-temporary standpoint is chosen with the full understanding that it is in part a cyclical choice and that it may need to be replaced in due course by its supplanted other. In other words, the development is not only upward in a spiral; it may also revert to an earlier stage of development, although its reversion may not be to the identity it previously possessed but will doubtless possess new characteristics. Thus the current age, the age of analysis in Western philosophy, has already reached its apogee and is on the decline. It is moving backward, so to speak, to a concern with metaphysical philosophy. From the standpoint of the *I-Ching*, such a movement in philosophy reflects the fact that change is a part of life and that change takes place when a viewpoint has reached its fullest development and needs to be replaced by the tendencies that it has suppressed.

In the ancient Ch'an and Zen Buddhist tradition, the "answers" provided by the sage were provided as prescriptions. In other words, skillful means or *upaya* were the grounds on which the answers were presented to students' questions. The same question from two different students might receive two opposite answers from the master. Indeed, the same question by the same student might be answered with opposite answers from the master. The answer chosen by the sage was chosen in terms of what the student was prepared to understand. Even the same student was at a different stage of development at different times. There was no simple right answer to the question. This is one reason the collected written sayings of masters sometimes appear self-contradictory. Each student (or the same student at different times) might need differing states of mind. This is why written works (at least of certain types of philosophy) always suffer from an inherent flaw.

On the other hand, one can learn from this approach. The new standpoint previously referred to may be taken as a dialectical and a natural development from the perspective of the *I-Ching*. From the standpoint of *upaya*, or Nietzsche's version of the philosopher as the physician, the new standpoint is not chosen wholly because it is the right or true standpoint but because it best fits the needs of the time or is the most appropriate medicine for the co-temporary cultural disease. It is interesting in this respect to contemplate that while Nietzsche saw that the philosopher most closely followed the practice of the physician, in ancient Chinese medical training, philosophy, particularly a philosophy that understood the dynamics of yin and yang, was an integral, if not key aspect of medical understanding. It should perhaps be pointed out that in Chinese medicine the ultimate condition of health was not considered to be a simple blend of yin and yang but rather one that reflected a proportion of 65 percent yin and 35 percent yang. This reflects an ultimate bias in medical priorities in favor of a condition that is less intense.

To put it in Western medical terms, it would be to favor low blood pressure over high blood pressure rather than an arithmetic mean between the two extremes. In philosophy, then, there is also a prescriptive dimension in the choice of the standpoints one ultimately favors and champions. This is the third element in the philosophical triptych of philosophical development. In terms of content, it could be said that it is more yin and therefore more feminine. Of course, this reflects the social needs of the time. But more than this it reflects the need to feminize, to replace conflict with care and combat with creativity. Ultimately, the beauty of the feminine is the lure to the production of something new, and the interaction of male and female is the dynamic process that gives meaning to life. It is this process and its importance that is illumined by highlighting the feminine. It is in this sense that it may be said that the proper prescriptive proportion between the sexes (with respect to balancing feminine and masculine poles within the human being) is 65 percent feminine and 35 percent masculine. In co-temporary society, it is obvious that the emphasis is clearly the reverse.

The feminizing of society is long overdue, though Lao Tzu certainly sounded this note a long time ago. (The proportion may vary with the differing needs of society.) Let us end this long-winded chatter with a practical example. In *The Analects* (*Lunyu*), Confucius makes a profound demarcation between humanity as the hallmark of a cultured person and religious practice or musical appreciation: The master said, "What can a man do with the rites who is not benevolent? What can a man do with music who is not benevolent?"[11]

In this passage, Confucius states that a person who possesses no humanity can possess no meaningful relationship to religion or music. It is evident from this statement that humanity is something that can and in fact must be possessed as a separate trait from ritual observance and musical appreciation. For the purpose of the current discussion, it will suffice to concentrate on Confucius's awareness of the fallacy of identifying an ethical life with a life of musical enjoyment. This answers the question of Mengele. Josef Mengele, the worst of the death-camp doctors in Nazi Germany, who cruelly experimented on inmates, was not merely an educated man; he read Dante and was a lover of Mozart. This gives rise to the question how a cultured man can be a human butcher? With a Confucian definition of culture, one who had not attained to morality would not be considered a cultured individual.

Mengele was not a civilized man according to Confucius. Mengele was a barbarian. A love of Mozart is not enough to constitute civilization.[12] Civilization requires the attribute of morality. Such is the contribution of Confucius. This is an application of the doctrine of the rectification of names.[13] One can borrow a methodological principle from Chinese philosophy to solve a problem of Western—nay, world—culture. Civilization can be restored to its ancient (Chinese) meaning, which includes, as a necessary component, a moral self. This ancient Chinese meaning may now be shared with the world as a whole such that civilization in name and deed can be returned to its true meaning. One may not, with legitimacy, refer to the calling for the death of all civilians of one race, nation, or religion as the expression of a moral view. To call a spade a spade, it is the expression of an immoral view, not a moral one.

It seems to be of enormous interest that by digging into the ancient philosophy of China, one may discover an ancient truth that can be applied to bring an important corrective to the twenty-first century. To create civilization, moral education is required. Without moral education, the prospect of repeating the barbarism of the previous century cannot be ruled out. Here is an example of how opening the West to China can create a benefit not only to the West but for the world as a whole.

Turning to Confucius to sort out a vexing problem in Western thinking is a good example of the value for the world today of coming into contact with new views, the fruitful dimension of dialectical philosophy. Of course, one can also comprehend this process as yang turning yin, as Western philosophy becoming more and more conscious of the need to turn to Eastern philosophy.

One can be educated in culture and civilization providing that education includes a moral education in addition to an intellectual, physical, and aesthetic education. This is not the last insight to be derived from an openness of both traditions, Chinese and Western, to each other, but it does display the advantages of openness and can therefore serve as incentive to shed the coils of a useless comparison of viewpoints and to engage in the active process of altering one's viewpoint to form a more humane and complete philosophical model for understanding and acting properly in the world today.

The ultimate lesson may be that one must remain in a continuous state of openness to the other—that is to say, to such views that show such openness in return. One cannot, in the world today, continue to regard the other as an alien other. Advances in transportation and communication have brought those who were previously considered alien others into instant contact. Technology has vastly outstripped moral advances. The cart is well in front of the man, not to speak of the horse; the rapid advances in technology have resulted in unforeseen moral crises. New moral dilemmas need solutions that are not provided by the technology that has produced them. Moral consciousness must transform itself swiftly to attempt to keep pace with technological advance.

The other is only another empirical instantiation of the self. The self and the other are wedded in a yin-yang harmony such that the self becomes the other and the other becomes the self. Yin-yang harmony is a model of collaboration, not one of conflict and competition. If cultures are perceived of as alien others destined for collisions and conflicts, then the outcome will most likely be a self-fulfilling prophecy. If cultures are perceived of as brothers, the prospects of harmony for the family of man will be great indeed. As is said in *The Analects*, "All within the Four Seas are his [the superior man's] brothers."[14]

CHAPTER 5

Justice and Confucian Harmony

Han Rui

A common feature of almost all social ideals, including Chinese and Western ideals, is that they regard harmony as of fundamental importance: a social ideal would not be an ideal if it did not aim at achieving some form of social harmony—that is, a harmonious relationship among all members of the society. But despite this, social ideals differ significantly from each other in their specific contents, a fact that also signifies striking differences between distinct philosophical traditions. This is especially the case with the Chinese and Western traditions. One primary difference between these two, perhaps, is concerned with the theme of justice. Ever since Socrates and Plato, justice has occupied the center stage of the Western tradition of political philosophy. The idea of justice is an integral part of various Western doctrines, from liberalism to individualism to democracy. This centrality of justice has found its contemporary manifestation in John Rawls's theory of justice, which, grounded in Kantian contractarianism, incorporates such fundamental Western values as liberty, equality, fairness, and democracy. By contrast, the theme of justice is never visible in the Chinese philosophical tradition. Confucianism, as well as other dominant strains of thought in China, has never invoked the theme of justice in the past thousands of years, let alone allocated it a central role in political thought. This difference in attitudes toward justice, which of course relates to deeper metaphysical, epistemological, and ethical underpinnings, may provide a good perspective for comparing the two traditions. And although skepticism about the commensurability of the two traditions is prevalent, this new perspective of comparison may inform us with insights into both traditions that have never been found before.

This essay, while proposing justice as a new angle for comparing the Chinese and Western philosophical traditions, has a very modest ambition. It will examine the main thoughts in both traditions, with the purpose of determining their respective conceptions of social harmony. After that the essay will raise a few tentative criticisms of Confucian harmony, as informed by the comparison of the two conceptions.

THE CONFUCIAN CONCEPTION OF HARMONY

Confucius's teachings, mostly in the form of conversations or exchanges with his disciples, are recorded in *Lunyu*, or *The Analects*. Despite his reputation, Confucius claimed that his teachings were not original but were only lessons transmitted from antiquity.[1] To him, the social and political order of the earlier Western Zhou period was ideal, and he therefore argued for its restoration. It is sensible to say that Confucius's social ideal models on the idealized peace or harmony of Western Zhou.

Confucianism is perhaps most marked for its emphasis on the moral cultivation of persons. Confucian ethics hold that virtues are necessary for a good life. An important concept is *junzi*, which means a person with moral or ethical nobility. Confucius thought that a *junzi* had *ren* (humanity or benevolence), *xiao* (filial piety), *yi* (righteousness), and *li* (in accordance with ceremonial ritual or general propriety), among other virtues. *Ren*, meaning "loving people," is the unifying theme of Confucianism. In a broad sense, *ren* connotes the complete or comprehensive ethical excellence of a *junzi*. It is for this reason that some contend that *ren* should be translated as "good" or "goodness."[2] However, at places in *The Analects*, *ren* is also explained in terms of caring for others. *Ren* in this narrow sense is often taken to mean "benevolence" or "compassion," which is just one virtue among others. *Xiao* means "filial piety." It requires children to show respect and obedience to their parents. Confucius considers filial piety significant for cultivating *ren* or moral excellence because the family is the first arena in which a person can practice love and respect for others. If a person cannot even love and care for his own parents, it is unlikely that he can have *ren* toward other people. *Yi* means "righteousness." It requires people to do what is morally right or appropriate in a particular context. *Yi* is often related to *li*, as what is right or appropriate is mostly hidden in the meanings of *li*—that is, rituals and ceremonies. *Li* connotes the ability to act according to ceremonial rituals or more generally to rules of propriety. A wide range of rituals or rules directs how people should behave in life, from ceremonies of worshiping ancestors to rules governing respectful and appropriate behavior between parents and children, men and women. By practicing *li*, an individual learns how to restrain himself and how to show courtesy or respect for others. Confucius stressed the importance of *li* because he thought that only by following *li* could a person cultivate his moral character and become a true *junzi*.

The emphasis Confucianism places on moral cultivation has to do with the fundamental Chinese worldview of that time. This view held that there was a universal order that was moral. Confucius believed that virtues or moral excellence would bring people into harmony with the *dao*, or "the way"—that is, the cosmic will of the impersonal universal order. Virtuous life, therefore, is a life in harmony with the *dao* and is a way of life that human beings ought to adopt. The ruler, as the son of heaven, rules with the "mandate of heaven," and if he wants his kingdom to last and prosper, he should also lead his kingdom to follow the *dao*. He should restrain himself and set a good moral example for his subjects. Confucius believed that moral self-cultivation of the ruler at the very top of the government was important because if the ruler does so, people beneath him will follow suit. In a conversation with Ji Kangzi, Confucius explained the power of the ruler's virtues with an analogy: "If your desire is for good, the people will be good. The moral character of the ruler is the wind; the moral character of those beneath him is the grass. When the wind blows, the grass bends."[3]

The Confucian political philosophy, therefore, consists in the belief that a ruler should learn self-discipline or self-restraint, should govern his subjects by his own example, and should treat them with love and concern. There are a few features of this tradition of political philosophy. First, Confucianism considers that rule by virtues is superior to rule by law. Confucius thought that the ruler's moral excellence was more efficacious than laws in educating and transforming people: "If the people be led by laws, and uniformity among them be sought by punishments, they will try to escape punishment and have no sense of shame. If they are led by virtue, and uniformity sought among them through the practice of ritual propriety, they will possess a sense of shame and come to you of their own accord."[4]

Under the rule of law, though people try to avoid breaking the law brazenly for fear of punishment, they may not acquire a sense of shame and may still commit crimes secretly. Under the rule of a virtuous ruler, people will be moved and educated by the ruler. By following the ruler's example to practice *li* and cultivate their moral characters, people will acquire a sense of shame and will not voluntarily commit crimes, whether openly or secretly.

The second feature of Confucian political philosophy is that it views the society as a family writ large. As mentioned earlier, the Confucian philosopher requires that the ruler should govern his subjects by his own example and, despite the powers he possesses, should treat them with love and concern. This preference of rule by virtues over rule by law is underpinned by the Confucian conception of society as family writ large. If the ruler resembles a parent and regards his subjects as his children, he will rely more on moral education or teachings than on punishments to persuade people to adopt a virtuous way of life. One implication of seeing society as a family writ large is that the Confucian society is hierarchical rather than equal, with people having higher or lower status. Just like unequal family relationships between the father and son or husband and wife, unequal social relationships exist between the ruler and ministers or between aristocrats and commoners, with social superiors having more authority than inferiors.

The third feature of Confucian political philosophy—and it relates directly to the previous one—is that it aggrandizes family values to the extent that they may even justify compromising the laws of society. Confucianism not only considers it acceptable if a person's behavior involves some unfairness or partiality due to loyalty to family, but it also considers that person inculpable. For example, Mencius is asked what the legendary sage-king Shun would have done if his father had killed a man. Mencius replies that the only thing to do would be to arrest him, for Shun could not interfere with the judge, who was acting on the law. But Mencius says that Shun would then have abdicated and fled with his father to the seacoast.[5] Absconding to avoid punishment, from an impartial point of view, shows unfairness and indifference to the victim, as well as connivance with his father's crime. Mencius, however, portrays it as honoring the value of loyalty to his family members. The story indicates the Confucian attitude toward impartiality: it is something that can be compromised, at least when it is in conflict with family values.

What then are the principles of good government according to Confucianism? Confucianism views society as a family writ large, with the ruler at the top of the social hierarchy caring for his subjects and teaching them with his own example. Given this conception of society, preserving the hierarchical social order and maintaining the good functioning of each part of it must be of great importance for the Confucian government. Thus, in Confucius's view, "Good government consists in the ruler being a ruler, the minister being a minister, the father being a father, and the son being a son."[6] This view indicates that when a person claims a title and participates in the various hierarchical relationships involved with that title, he should live up to the meaning of that title. That is, he should try to fulfill all the responsibilities and obligations that the title prescribes and should show respect to his superior as well as care for his inferior. Confucius thought that society had become decadent because people did not fulfill what their titles required them to do. Confucius's analysis of the lack of connection between actualities and people's titles, and his emphasis on the need to correct such circumstances, led him to his theory of *zhengming*. As Confucius says to his disciple Zilu, the first thing he would do in undertaking the administration of a state is *zhengming*, which is to rectify the behavior of people so that it matches the prescriptions of their titles. When people live up to their titles, doing what they should do and only what their titles require them to do, their society is in good order and therefore in harmony.

It is appropriate now to describe the kind of social harmony that Confucian political philosophy envisages. A Confucian ideal society is perhaps most characterized by a hierarchical structure modeled on the family. The ruler in this ideal is marked by his possession of moral excellence. The moral power of the virtuous ruler allows him to win a following without recourse to physical force, and the society will be in good order if the subjects follow the example of their ruler to adopt a virtuous way of life. Good government, then, consists in parties at different levels in the social hierarchy doing what their positions require them to do without overstepping the mark. The Confucian three cardinal guides, for example, require that the ruler guide the minister, the father guide the son, and the husband guide the wife. This social ideal requires individuals to fit themselves into the hierarchy, being loyal and obedient to their superiors and caring for their inferiors. If everyone conforms to this hierarchy and fulfills what their titles require them to do, just like the example set by the ruler at the top, then the whole society is in good order. Confucian harmony is therefore a harmony with unequal social relationships between individuals, realized when individuals accept their own higher or lower status in the hierarchy.

THE WESTERN LIBERAL CONCEPTION OF HARMONY

Confucian political philosophy holds that social harmony consists of each person depreciating his own interests and obeying his superiors in the social hierarchy. This means that if his own interests are in conflict with those of the family, community, or society, he may have to sacrifice his interests, whether doing so is just or unjust to him. This attitude, however, is certainly unacceptable in Western liberal political philosophy, which centers on guaranteeing the legitimate interests of each individual.

Western liberal moralities center on individuals. Although there are various moral philosophies within the Western tradition, many in the West believe that morality is best seen as consisting of the set of characteristics associated with Kant's moral philosophy—the assumption of individuals as equal, rational, and autonomous agents and a belief in universal laws validated by pure reason.[7] Kant's "categorical imperative" requires that we act only on precepts that we can will as universal laws. The moral precepts, derived from reasoning between equal, rational, and autonomous parties, are universal because they are symmetrical: if a precept can be justified to one party, it can also be justified to another. The morality validated in this way is like a social contract, an outcome of consensual agreement between parties. This morality certainly implies an equal concern for each and every individual, for a moral precept will not be acceptable if it can be justified only to some but not to others. This equal concern for individuals is central to Western philosophical conventions, from liberalism to individualism to democracy.

In the Western tradition, the idea of equal concern for individuals is closely related to justice. Aristotle said that justice is a kind of equality. The value of equality means that all people have equal moral worth or enjoy equal moral status. The closeness between justice and equality is based on the belief that justice obtains only when all individuals have been given equal concern, although it is a profound question what really counts as equal concern. For Aristotle, justice or equality involves treating equals equally and nonequals nonequally. Classical liberalism holds that equal concern means that individuals have equal political rights and civil liberties, and an equal opportunity to compete in the free market. The philosophy of democracy interprets equal concern as the equal importance of individuals in shaping social institutions and policies that will have an influence on their own lives. It is therefore sensible to say that the value of equality is essential to the various strains of thought in the Western individual-oriented tradition, despite the fact that each strain gives it its own interpretation.

The central role of justice or equality in constructing a well-ordered society has perhaps been best articulated by John Rawls. At the beginning of his landmark book *A Theory of Justice,* Rawls claims that "Justice is the first virtue of social institutions."[8] No matter how efficient or well arranged social institutions are, they must be reformed or abolished if they are unjust. The primary subject of Rawls's justice is the basic structure of a society. In Rawls's view, the basic structure has a profound influence on the life prospects of all individuals because it assigns rights and duties and distributes benefits and burdens of social cooperation among people. For Rawls, a society is just only when its basic structure grants equal concern to all members by distributing among them, in a fair way, such primary goods as liberties, rights, opportunities, income, and wealth.

There are at least two aspects in which the Rawlsian conception of justice ensures equality among individuals. First, Rawls's two principles of justice are derived from pure reasoning of free and equal parties in a hypothetical contractarian situation—that is, the original position.[9] The parties in the original position are equal in the sense that all have the same rights in the procedure for choosing principles: each can make proposals, submit reasons for their acceptance, and so on. Obviously, the purpose of these conditions is to represent equality between human beings as moral persons. To ensure further equality in their reasoning, Rawls situates the parties behind a veil of ignorance, which deprives the parties of certain kinds of information, such as their social status, intelligence, strength, conception of good, and plan of life, to prevent them from favoring their own cases in choosing the principles of justice. The original situation, with equally situated parties behind a veil of ignorance, is intended to guarantee that whichever principles are chosen, they will be equally justifiable and therefore acceptable to all parties. The Rawlsian justice yielded by the original position therefore is committed to granting equal concern to individuals, since it is a conception of justice to which all free, equal, rational parties behind the veil of ignorance will consent.

Derived from the original position constituted by equal parties, the Rawlsian justice is also characterized by an expansion of equality from the political arena to the economic arena.[10] Rawls's first principle of justice stipulates that each person is entitled to equal rights and basic liberties compatible with those of others. Since equal entitlement to these basic political rights and liberties has long been recognized and accepted by the Western liberal tradition, there is little controversy over the first principle.

Rawls's second principle is concerned with mitigating inequalities in the distribution of income and wealth that are caused by morally arbitrary factors. This principle is comprised of two parts. Part one requires a fair equality of opportunity, which consists of opening positions or offices to all talents and providing public education to ensure that all children stand on the same starting point for competition. Part two is often referred to as the difference principle, which stipulates that wealth inequalities should be reduced, but only to a point where they can still preserve efficiency of economy. Arguments supporting the difference principle relate to the moral arbitrariness of people's natural endowments, which Rawls thinks accounts for wealth inequalities when the fair equality of opportunity obtains. If it is unfair that morally arbitrary factors should influence the distribution of wealth, then natural endowments also should not be allowed to affect the distribution. But the difference principle stipulates only a mitigation, not a nullification, of economic inequality, allowing some inequalities on the ground that they are necessary to maintain economic efficiency, the growth of which benefits all people, especially the disadvantaged in society. The Rawlsian conception of justice therefore goes further than other doctrines in trying to guarantee an equal concern for individuals. For Rawls, justice or equality involves not only equal political rights and liberties but also relatively more equal distribution of income and wealth among individuals. His core argument for reducing economic inequalities, which relates the moral arbitrariness of certain affecting factors to the unfairness of wealth inequalities, has been influential. He was by no means the first liberal to emphasize the importance of social justice, but his *A Theory of Justice* ignited numerous discussions about distributive justice or social justice.

It must also be mentioned, for the purpose of this essay, that Rawls's theory of justice is proposed as an alternative to utilitarianism, whether classical or average utilitarianism.[11] The utilitarian concept of justice holds that social institutions should be arranged so as to achieve the greatest total utility (that is, satisfaction of rational desire) summed over all individuals, or the greatest average utility for each individual. Since utilitarianism takes the principle of rational choice for one man as the principle of social choice, Rawls thinks that it fails to take seriously the distinction between persons. Rawls argues that utilitarianism could imply sacrifice of individuals for any possible increment of total or average utility. For Rawls, however, justice consists precisely in equal concern for individuals and must prioritize equal concern over maximization of utility.

From the above survey of some major Western philosophies, we can perhaps abstract a Western conception of harmony. The Western liberal tradition is evidently an individual-oriented one. It relates justice to equality, committed to the belief that justice consists in equal concern for individuals. In contrast to the Confucian harmony, which is a top-to-bottom, society-oriented ideal, Western liberal harmony is a bottom-to-top, individual-oriented ideal. It starts from the equal and fair treatment of each and every individual and ends, ideally, with the whole society having little conflict and being harmonious.

SOME TENTATIVE CRITICISMS OF CONFUCIAN HARMONY

The previous two sections have abstracted conceptions of harmony envisaged by the Chinese and Western philosophical traditions. This last section will attempt to raise a few criticisms of the Confucian ideal and will also consider the challenge that Confucianism deals at the Western way of thinking.

The first criticism is concerned with doubts whether the Confucian ideal can be realized by the means it proposes. As we know, Confucian harmony is a top-to-bottom ideal, with a virtuous ruler regulating the whole hierarchical society beneath him with his moral power. Coercive laws are supplementary, and when the ruler possesses virtues, the whole coercive apparatus of law and state controlled by the ruler will assist the rule of virtues. While this seems to be a desirable ideal, the methods Confucianism provides for realizing the ideal fall short. In the first place, Confucianism offers little guidance as to how to ensure that a person with virtues is on the throne. The Confucian classics mentioned the legendary sage-kings Yao and Shun, who abdicated and handed over the throne to a successor chosen for his moral excellence. But for thousands of years in feudal China, thrones were either hereditary or obtained through force. A hereditary ruler might not possess the required moral excellence even if his predecessor did. Thrones established by force were also problematic: while some were the outcomes of righteous rebellions against decadent and corrupt rulers, others were marked by usurpation, which involves bloody conspiracy and killings within the royal family. Since there is no guarantee that the person succeeding to the throne will be one with moral excellence, there is strong reason why Confucius so stresses the moral cultivation of the ruler. The teaching of virtues, especially through *li*, or rituals, is to help the ruler understand the profound meaning of morality so that he can lead his kingdom toward peace though complying with the *dao* of the universe. But the thousands-year-old feudal history of China has shown that moral cultivation of the ruler is not always successful. While some rulers did follow the Confucian teachings of moral cultivation and rule of virtues, many others did not. And since they were under little restraint and had absolute control of the coercive legal and political apparatus, their rule could easily deteriorate into dictatorship, authoritarianism, or oligarchy. Moral excellence on the part of the ruler is central to a Confucian political ideal, but Confucianism does not offer a sure way to guarantee either a virtuous person being selected as ruler or a successful moral cultivation of the ruler.

A more important criticism of the Confucian ideal arises from the concern for justice. Following Rawls, the theme of justice is taken to involve not only the granting of equal political rights and liberties but also the mitigation of economic inequalities. Both aspects of justice aim to recognize equal moral worth in individuals. Confucianism, however, fails to incorporate justice into its ideal. We will present this criticism in terms of the two aspects of justice respectively.

First, due to its lack of concepts of rights and liberties, Confucianism has been frequently criticized for failing to provide adequate protection of individuals. Individuals may have to sacrifice their own legitimate interests if they are in conflict with those of family, community, or society. For example, it was common practice in China that a family would marry a child into another family to enhance the power or status of the family, without considering the child's own will. Without the protection of freedom of speech, a minister admonishing his ruler would sometimes face punishment. The case was even worse for dissenting individuals. They often had to risk their lives to point out abuses of power or just plain bad thinking by authorities. It might be true that some germs of rights and liberties are implicit in the Confucian classics. Mencius, for example, advised kings to attach more weight to the opinions of his people than to those of his ministers and officers in making certain crucial decisions.[12] Xunzi recognized the need for subordinates to speak their views freely to their superiors.[13] These ideas may hint at the need to protect a space in which individuals may speak freely without fear of suppression and may entail the right to exercise certain capacities when one has justifiable claims on others. But these implicit ideas have never been developed or institutionalized into concepts like rights or liberty.

A relevant criticism is that the dignity of the individual cannot be honored without recognition of individual rights and liberties. In classical liberalism and individualism, the individual is viewed as surrounded by a protective cordon of rights, which define his or her freedom. Rights in this sense honor the individual's dignity with the affirmation that he can do something without being interfered with by others and that he can refuse unfair claims or requirements imposed on him. Since Confucianism does not have such concepts as rights or liberties, individuals have no way to defend themselves on fair terms but are left with no alternative but to succumb to the powerful. It is therefore not surprising that people in a Confucian society tend to be either subservient to superiors or peremptory to subordinates.

Some have held that it is possible that the Confucian tradition incorporates the concept of rights in some sense. To render rights compatible with Confucianism, it is also possible that such rights as incorporated are not grounded in the idea of the independent moral worth of the autonomous individual. Some try to justify rights in terms of community interests. For example, it has been argued that rights in the sense of proper claims to be protected in one's speech, even when protesting and dissenting against authority, can be justified as conducive to the health of the community.[14] Along this line of thought, some also suggest that rights may be seen as a fallback that is necessary for protecting individuals' interests when caring relationships between individuals irretrievably break down.[15] These efforts seek to remedy the insufficiencies of the Confucian ideal caused by a lack of protection of individual rights.

Confucians do have a reply to this criticism. The Confucian view of society as the family writ large imposes a great responsibility on the ruler, or the ruling class, requiring rulers to treat subjects like their own children. Some argue that the Confucian framework of responsibilities to others can afford significant protection of the individual and even address the human need for community and belonging better than rights frameworks.[16] But can the Confucian responsibility frameworks provide unfailing protection of dissenting individuals? While the ruler may take up this responsibility and love his subjects in the same way that a parent loves his children, there is no institutional guarantee that he will fulfill this responsibility, especially when the dissenting individual seriously threatens communal or social interests. As history often shows, the lack of protection of the individual is especially a problem when the ruling class is not composed of people who possess the required moral excellence. In that case, the ruling class easily turns into one whose chief concern is to maintain its privileges and authority and to aggrandize its interests as much as the ruled can stand. With the Confucian teachings of self-restraint and subservience, the danger is that people tolerate unfair circumstances and behaviors that imperil society. Superficially, this status quo may still be harmonious, as the ruled will be more conforming than rebellious. But it would be strange to say that this is a real harmony worth pursuing.

Even if the ruler desires to fulfill his responsibility for his subjects, the Confucian responsibility framework may still be criticized for its obvious commitment to paternalism. People have different life plans, conceptions of the good, and priorities of values, and therefore they may not agree with all decisions made by the ruler concerning their lives. In a Confucian hierarchy, however, even if individuals know what is best for their own lives, they cannot insist on their own points of view; rather they must accept the ruler's authority in making decisions for them, under the assumption that the ruler knows what is best for all. In sum, paternalism is intrinsic to Confucianism and simply does not accept the idea, so fundamental to modern liberal democracies, that people have the right (within the limits of the law) to choose the kind of life they think is good.

The justice-centered criticism of Confucianism applies not only to political rights and liberties but also to wealth distribution. In the main, Confucianism is unconcerned with wealth inequalities in society. It supports the idea that since the ruler was the son of heaven, all lands belonged to him. The ruler might also grant land to feudal aristocrats (the practice of enfeoffment). They would allow common people to plant on their land, but only if they paid tributes or taxes. It is true that Mencius advocated a *well-field land system*, according to which land was divided into nine parts, with eight being "private" and one being "public." All peasants had to plant the public lands before working on their respective private lands. But no peasant could really own a plot of land: the terms *public* and *private* meant only that the yields of the public land belonged to the landlords while those of the private land belonged to the peasants. But it never seems to have been an issue, under the Confucian framework, whether the ownership of land and the division of labor were fair or whether the gravely unequal distribution of wealth in society was just. While Confucianism does emphasize compassion, benevolence, and responsibility, it does not translate these concepts into concrete legal or political instruments to which individuals can resort when the ruling class fails to answer the call of virtues.

It is perhaps sensible to conclude that the Confucian conception of harmony consists of the avoidance of social conflicts at the expense of individuals. It seeks to unify the whole society with a hierarchical morality, teaching the ruler to learn self-restraint and the people not to go against their social superiors. As a society-oriented ideal, it is ready to sacrifice individual interests or to suppress individual dissenting voices whenever they threaten the harmony of the whole society. Confucianism protects neither the political rights and liberties of individuals nor their fair requirements of lesser wealth inequalities. In light of these deficiencies, it is reasonable to say that Confucianism does not seek justice for individuals. Confucianism may contribute to a kind of harmony, but it is a form of harmony without justice, at least as it is understood in the Rawlsian sense of fairness.

Given these criticisms, it is also worth considering a common Confucian criticism to the Western liberal conception of harmony. This is the argument that Western individual-oriented morality ignores the social nature of human beings and thus, so the argument goes, contributes to an excessively atomistic or individualist conception of the person.[17] From the perspective of Confucianism, the Western liberal focus on the individual may predispose people to isolate themselves from members of their family, community, or society. Thus each lives in his own little world in solitude. While this is a serious criticism, it tends to oversimplify Western liberal individual-oriented ethical thought. A deeper look into the Western tradition would reveal that the social nature of persons is not denied by all Western liberal thought. For example, while Hobbes assumes an atomistic view of human beings (though he is not overtly liberal), both Rousseau and Locke seem to require no such view. Rousseau builds his social philosophy around the general will, while Locke, who certainly defends private property, also claims that people are inherently reasonable and capable of forming collective associations on the basis of shared interests. While the atomistic view is closely related to a negative concept of liberty—that is, absence of interference from others—it does not sit so well with a more positive concept of liberty—that is, a version of liberty that emphasizes the ability to act in such a way as to take control of one's own life.[18] Unlike negative liberty, which is usually attributed to individuality, positive liberty is sometimes attributed to collectivities or to individuals considered primarily members of given collectivities.

John Dewey's philosophy can perhaps serve as a good reference point as it explains why the Western individual-oriented tradition may not lead to an atomistic picture of human life. In his earlier works, he worked out an idealist critique of classical liberal individualism.[19] Dewey rejected the classical liberalist view that the individual was an independent entity in competition with other individuals and that social and political life was a sphere in which competition and conflicts of private interest took place. Instead, he sought to view individuals as social and relational; he said that individuality involves participation in shaping the conditions of the social organism of which each is an integral part. Dewey's understanding of freedom was therefore also "positive." Freedom was not, for him, the absence of external constraints; rather it was the positive fact of participation in social life. As Dewey put it, "Men are not isolated non-social atoms, but are men only when in intrinsic relations to one another, and the state in turn only represents them 'so far as they have become organically related to one another, or are possessed of unity of purpose and interest.'"[20]

Dewey's stress on the social nature of individuals, however, does not commit him to a denial of individual-oriented morality. Rather, he is anti-elitist and an advocate of democracy. He is against the idea that the wise few can discern the public interest. He argues for democracy because he believes that democracy, in the way he understands it, is important for achieving a high level of social harmony. Dewey proposes three lines of argument for democracy. First, democracy allows individuals to express their interests and to protect themselves from experts or elites concerning where the interests of the people lie.[21] Democratic discussion or consultation can help inform people of real social needs. Second, democracy is also a form of social inquiry, a way of dealing with conflict of interests in society. Democratic politics is not simply a channel through which individuals can express and assert their interests but a forum in which individuals can find out what their interests are, whether they are in conflict with those of others, and how to solve the conflicts.[22] Finally, democracy for Dewey is required for the individual's freedom. His positive freedom theory entails that individuals are free in the sense that they play a role in shaping the society to which they belong.[23] Of course, Dewey's conception of democracy goes further than the common view of democracy as a set of specific political procedures and institutions. But he believes that through public discussion and communication, a society can achieve a high level of harmony among its members.

Though Dewey's idea of democratic harmony may be as idealistic as Confucian harmony, it represents a defense of the Western individual-oriented tradition. The Western liberal conception of harmony does not seek to subject some to others, suppress dissenting views, or sacrifice individual interest for the sake of community interests. As Dewey's idea of democratic harmony illustrates, it emphasizes that individuals best protect their own interests, realize their potentials, solve problems and conflicts through public discussion, and thus shape the common good through public discussion. We must say that this democratic harmony is just since it gives equal concern to each individual, whether in political or economic terms. While Dewey provides only one version of Western liberalism, I think it fair to say that all versions of Western liberalism emphasize justice and equality (though the nature of equality may vary from a more formalist understanding of the term, as in Locke or Hayek, or from a more social and substantive understanding, as in Rawls or Dewey), and thus Western liberal conceptions of harmony are inconceivable without some notion of justice being intrinsic to a diverse set of interests forming a social totality. By contrast, Confucian conceptions of harmony may be attained without equality or justice.

CHAPTER 6

Two Concepts of Order: An Essay on Harmony and Order versus Spontaneity and Revolt in Western Thought

Hélène Landemore

Every human group faces the problem of how to reconcile the tension between, on the one hand, the requirement of public order and the need for a coordinated and cooperative life in common with others and, on the other, the expression of individual needs, desires, and freedoms, which may sometimes conflict with each other. The Western philosophical tradition has typically conceptualized this tension as one between the collective goal of political order and the individual value of freedom as "negative liberty," which may or may not support a right of revolt against an order perceived as unjust.

The point of this essay is to explore a set of moments or conceptual breakthroughs within this Western tradition that demonstrate one way of partially resolving the tension. In this paper I am thus interested in exploring the concepts of order and harmony versus spontaneity and revolt from within the Western tradition, with which I am most familiar, in an effort to start a dialogue between two apparently different—some would say incommensurate—philosophical worldviews. I will take the contrasts and nuanced relationships between these four concepts as the starting point of a larger reflection about the difficulty of conceptualizing social life as harmonious without silencing individual voices or blocking the possibility of change.*¹* I will more specifically attempt to identify moments in Western thought that have allowed for a transformation in the Western understanding of political order and social harmony, from one that implies the rigid control of individual spontaneity and the impossibility of revolt to a modern (liberal and democratic) concept of order that encourages spontaneity, feeds on it to a degree, and attempts to institutionalize revolt such that it stabilizes rather than upsets the system. My hope is to offer a clear picture of the Western conceptual solution to what is, in many respects, a universal problem and to do so in terms accessible to both the Western and Eastern traditions. Hopefully this picture will be useful in helping to identify a comparable solution, or a contrasting one, in the Eastern tradition. From there, the possibilities are endless. They include, among others, a comparative approach to both worldviews, drawing out clear parallels between various authors and schools from both traditions, or perhaps the "invention" of new paradigms within Western and Eastern philosophies on the basis on insights borrowed from one another.

Let me begin by emphasizing a particular contrast that needs explanation. Arguably, most premodern Western philosophers—from Plato to the late medieval thinkers—embraced the identification of political order with both a socially harmonious whole and a rather rigid hierarchical organization. The whole was harmonious to the extent that the parts in it—different groups of people—played their roles, and only their roles, in a relatively unchanging hierarchy. At the top of that hierarchy were the rulers, whether kings, masters, or lords. At the bottom were the ruled—producers, peasants, and the lowest of the low, slaves and serfs. Somewhere in the middle were the intermediary categories of the military and the clergy. This ideal of order offered, in theory at least, very little room for individual freedom as we today conceive of it, let alone the possibility of revolt.

In practice, of course, things were not as tidy. Aristotle observed (and deplored) the natural entropy that makes order turn into disorder and well-functioning regimes become corrupted over time. Further, even the most perfectly implemented, rigid order must have had to deal with the inevitable minor and less minor rebellions of individuals resisting the system. Nonetheless, it is generally acknowledged by historians and political theorists, such as Benjamin Constant, that the "liberty of the Ancients"—the liberty of citizens in republican, precommercial states, usually at war with one another—was the liberty of individuals as citizens, not so much the "modern liberty" of individuals as private individuals. Such liberty found its expression in the fulfillment of an impersonal duty, not in the satisfaction of idiosyncratic preferences or the pursuit of spontaneous desires. In the ideal governing the ancient world, therefore, not only was individual spontaneity very much controlled, but each class was assigned a function. Being born in a class was, in theory at least, a life sentence to fulfilling that function and no more (and no less). That's because the possibility of rebelling against such an order—the possibility of deviation, change, and indeed revolt—was not built in the ideal.[2]

Today, by contrast, in the Western world, dissenters, conscientious objectors, norm breakers of all kinds are not only tolerated; they are celebrated. Modern liberal societies value dissent, for the most part as a source of social energy and creativity. The expression of individual spontaneity and disagreement is seen (again at least up to a point and within limits) not as a social disease threatening the life of the body politic but as a vaccine that will make it stronger. Artists—the norm breakers par excellence—are thus sometimes as influential these days as some heads of state (think Hollywood actors and rock stars). As to revolt, it has now been elevated to a James Dean attitude, by which anything from haircuts to slogans on T-shirts can count not just as an expression of individuality but as a form of rebellion against the system. Perhaps as a result, because liberal societies offer so many outlets for feelings of frustration and anger and for expressions of dissent and difference, actual outbursts of violence are fairly rare—not because of the repression that would ensue but because things do not need to get so out of hand for things to change. What liberal societies have, occasionally, are public demonstrations. Those, however, are not meant to overthrow the whole political order but merely to shake it, let off some steam, and induce internal reforms.

How did the West move from one worldview to the other? Has the East made that move, and if so, how does it compare to the Western transition? Is there anything one side can learn from the other?[3] I will limit myself here to answering the first question by pinpointing two key breakthroughs in the history of Western thought: first, the initial, liberal reconciliation of order and spontaneity within the social contract tradition of the seventeenth and eighteenth centuries, which replaced the priority of the collective with the priority of the individual; and second, the linkage between order and harmony, on the one hand, and free discussion, deliberation, and unregulated spontaneity, on the other. Though the first attempt at reconciling order and harmony with spontaneity and revolt would turn out to be unstable and unsatisfying, for reasons I discuss below, social contract theorists like Hobbes, Locke, and Rousseau made it possible to conceptualize spontaneity, freedom, and even revolt as legitimate elements of a theory of political order, and conversely they put the burden of proof—the need for justification—on any form of political, institutionalized order. Eventually, this opened up the necessary space for thinkers of the nineteenth and twentieth centuries—beginning with John Stuart Mill, followed by pragmatist philosopher John Dewey and economist Friedrich Hayek—to develop a more nuanced account of the relationship between social harmony and individual spontaneity. As I will show, out of Mill's famous defense of freedom of thought and expression and his argument for representative government as "government by discussion" come both Dewey's portrait of democracy as social inquiry, in which truth and harmony depend on deliberation between heterogeneous views, and Hayek's model of self-organizing spontaneous orders, exemplified in the market. I will argue that these two branches stemming from Mill are not only the culmination of the liberal tradition opened up by the social contract theorists but also represent the great achievements of modern Western thought.[4]

The first part of this paper offers some general reflections about the concepts of order and harmony versus spontaneity and revolt as interpreted within Western thought, drawing out some of the more obvious contrasts and oppositions. The second part traces the first major breakthrough in reconciling these opposite elements to the social contract theorists, who develop the idea of a political order as first and foremost a guarantee for individual liberties. The third part connects the Deweyan tradition of deliberative democracy and the Hayekian defense of spontaneous orders to Mill's arguments about liberty and government by discussion.

A FEW REFLECTIONS ON THE DUALISM HARMONY AND ORDER VERSUS SPONTANEITY AND REVOLT

I suggested in the introduction that the contrast between harmony and order on the one hand and spontaneity and revolt on the other was political, comparable to the classical Western opposition between law and freedom, positive and negative liberty, or even law and justice. Neither harmony nor spontaneity, however, are political concepts per se. *Harmony*, which means "agreement" and "concord" in ancient Greek, is first and foremost a musical and, more broadly, an aesthetic concept, characterizing a certain combination of elements—notes, colors, dimensions—that induces a pleasing feeling in the observer.

Applied to the political world, *harmony* suggests a polity in which different classes live in peace with each other and agree on the ends of the community, and in which a sense of respect and civility permeates social interactions. Plato's ideal republic is the paradigm of such a social order. For Plato, the most harmonious polity was that in which philosopher-kings ruled, whereas the most disharmonious and unhealthy polity was a democracy, where all sense of hierarchy was lost, resulting in a regime characterized by a cacophony of desires, which he also compares to a costume of jarring colors.[5]

The concept of spontaneity is not directly opposed to those of order and harmony, but it has a clearly upsetting potential for them. Whereas order and harmony belong to the world of culture, spontaneity characterizes the self-generating elements of nature, like the blooming of a flower or the unconstrained manners of a child. Spontaneity more generally characterizes anything that does not require application, planning, and labor.[6] The closest political concept with which spontaneity can probably be associated is that of negative liberty, or "freedom from" any form of constraints, particularly those imposed by society.

Harmony and spontaneity can be at odds. Harmony is itself a kind of order, and like order it is a relational concept that evokes an arrangement between parts (tunes or proportions). Spontaneity is to harmony what the improvisations of a jazz band are to the music of Rameau or Bach. Yet while there may be something disturbing about the unpredictability of jazz, the disharmony need not be grating. The spontaneity can be playful, agreeably surprising, exciting.

Unlike "harmony and spontaneity," the conceptual couple "order and revolt" is, I believe, explicitly Western, political, resolutely antithetical, and much grimmer overall.[7] There exist many definitions of the concept of order, but for our purposes the following ones will suffice. Order can be:

1. A condition of methodical or prescribed arrangement among component parts such that proper functioning or appearance is achieved

2. The established system of social organization

3. A sequence or arrangement of successive things

4. The prescribed form or customary procedure [8]

The common point of these definitions is to emphasize a static, sometimes preexisting arrangement of component parts that guarantees the proper functioning or appearance of the whole. This order can be an object, a time sequence, a form, or a procedure. While it shares with the concept of harmony the idea of a whole organized according to certain rules, order is far from a musical notion. Rather, order evokes the austerity of law and justice. It also has the threatening face of those who guarantee that order: the army, the police, judges.

Spontaneity and revolt, by contrast, suggest life, chaos, and disruption. Revolt in particular screams the threat of revolution and even anarchy—better no order than *that* order. Thus revolt is a more menacing concept and more antagonistic to the idea of order, even when it actually paves the way for, and demands, a new kind of order. On the other hand, whereas revolt is a temporary aberration, a moment of violence that is not sustainable in the long term, spontaneity is consubstantial to any form of life and cannot be eradicated without eradicating life itself. Spontaneity is a repetition, where revolt is a transition. Revolt need not be fatal for any existing order. Spontaneity is something that must be accommodated in some form or another.

So far we have associated order with a form of political or social organization and opposed it to individual spontaneity, thus suggesting a broader opposition between nature and culture or nature and human organization. One of the great discoveries of the social sciences, however, is that not all human orders are artificial and intentional and that many orders can be characterized, however paradoxical this may sound, as "spontaneous." Nature contains many orders that have not been planned or designed, such as crystals or the ways in which irons fall within a magnetic field. Human orders themselves, such as the family, the clan, the market, need not be the deliberate design of anyone in particular. They are rather the unplanned result of human interactions that, over time, have ended up crystallizing in certain predictable patterns, which one may call laws (in the sense of common law, not positive law) or norms. These laws and norms may be crushing for the individual, such as the norm of genital mutilation or honor killing; alternatively they may be enabling, such as the laws and norms of the market. An order need not be an organization—that is, an arrangement planned by anyone in particular— to deserve the name; rather, to be an order, it must allow for a certain amount of predictability. We will see in section three of this paper how Hayek uses this discovery to conceptualize a new kind of human order. At this point, it is enough to observe that there is more than one way to conceptualize the relationship between order and harmony on the one hand, and spontaneity and revolt on the other.

Now that we have drawn the most obvious similarities and oppositions between these concepts, we can formulate the political question that underlies them: How do we create a social world characterized by harmonious relationships—not just between rulers and ruled but among the ruled themselves—without crushing spontaneity and creativity? Can we create the conditions for that harmonious order to evolve over time without breaking apart? In short, can there be any form of harmonious order that is not oppressive and repressive to individual spontaneity?[9]

I will now focus on the first moment of transition in Western thought that made it possible to think of spontaneity and revolt together with order and harmony. This first moment, I argue, is to be found in the writings of the social contract theorists, expressed as an attempt to confront the duality of (positive) law and (negative) freedom.

THE SOCIAL CONTRACT THEORY MOMENT

The philosophical break in Western thought that allowed for thinking beyond the radical opposition between political justice as harmonious order and individual freedom as lawless license occurred with social contract theory. Social contract theorists were aware of the contradiction between order and liberty, and they strove to offer a conceptual reconciliation balancing the two.

Social contract theorists started from the liberal premise that social orders are for the sake of individuals and not the other way around. They rejected the Aristotelian view of man as a political animal and posited instead natural individual rights; political communities thus are artificial constructs into which human beings enter only by (tacit or explicit) consent. With this conceptual revolution under way, it became necessary to carve out a space in theories of political authority for individual freedom and even the possibility of revolt.

From that perspective, social contract theorists marked a radical departure from the ancient world by shifting the burden of proof onto all existing political authorities. They asked the question: What makes *this* political order legitimate? Their answer was that a given, de facto political authority is legitimate only to the extent that it stems from the consent of the subjects and fulfills the function for which it was created in the first place: at minimum, preserving citizens from the threat of violent death, as in Hobbes; at best, implementing the general will, as in Rousseau; and, somewhere in between, safeguarding subjects' property (including their lives), as in Locke. For the social contract theorists, political order was now for the sake of individuals, not the other way around.

Once political authority is legitimized, however—once the social contract has been entered—the degree to which political authority is supposed to make room for individual freedom, spontaneity, and even revolt can vary quite drastically. In Hobbes, for example, the Leviathan provides a stable order, for the sake of which individuals at war with each other in a chaotic state of nature surrender their natural rights and liberties, including the right to render justice themselves.[10] What Hobbesian individuals lose in natural freedom—the spontaneity to do whatever pleases them, including killing others—they gain in civil freedom, which is the liberty to do what the law does not forbid. This liberty can be, in fact, fairly extensive. The problem is that nothing protects subjects from the occasional arbitrariness of the sovereign. Because the law is reduced to its most positive expression—justice is what the sovereign says it is—there is no appeal beyond it. Because there is no justice but what the sovereign speaks, it is impossible to denounce as unjust the commands of a ruler, even a despotic, arbitrary one.[11] In other words, if the sovereign decides that the life, liberty, or property of a citizen needs to be taken to serve some interests of the state (provided this is genuinely an interest of the state that can be justified in rational terms and that conforms to the laws of nature), he is entitled to them all.[12] In Hobbes, therefore, room for spontaneity is precarious. In any case, it certainly does not include the right of revolt. Revolt would be indeed tantamount to reclaiming the right to render justice yourself, which was surrendered in the social contract. In truth, the only right that individuals retain under Leviathan is a right to flee from their executioners, or at least the right to try to do so and not to be punished for trying.[13]

In Locke, the situation is less dire because the social contract is not based primarily on the fear of violent death but on a desire to avoid the "inconveniencies" of an otherwise relatively peaceful state of nature.[14] As in Hobbes, the government is legitimate to the extent that it has elicited the tacit, if not explicit consent of the governed. But unlike Hobbes, Locke acknowledges that even an initially legitimate government can ossify into a tyranny and turn lethal to the individual rights it was meant to secure. One particularly sacred limit is the property of subjects. While the government is entitled to send soldiers to a certain death,[15] nothing can justify taking a penny from any subject without his consent.[16] The Lockean view of the political order is not that it secures life only, as in Hobbes, but that it creates the condition for the safe enjoyment of one's property. In that respect any arbitrary act on the part of the government violates the social contract and entitles individuals to rebel against the government authority. Individuals retain this right unconditionally because, unlike for Hobbes, subjects do not abdicate their judgment of justice to the sovereign. They are entitled to interpret the laws of nature in their own ways, and no one but God can say if they've interpreted them rightly or wrongly. That said, Locke also acknowledged that if the violation does not amount to a series of offenses, individuals are more likely to make a mistake in their assessment of injustice. So in practice for Locke, the right to overthrow the government arises only in cases where the people had to suffer "a long train of abuses, prevarications, and artifices, all tending the same way."[17]

In Rousseau, the reconciliation between order and freedom is ensured by definitional fiat. In Rousseau's view, the general will is said to be oriented toward the common good, and thus it cannot err. Whatever it wills, citizens ought to will it too. Consequently, coercion of all according to that general will is not coercion of any one.[18] Civil freedom is not just what one is at liberty to do in the silence of the law, as in Hobbes; it is obedience to that very law. Freedom is "positive" in Isaiah Berlin's sense of the term:[19] it is an empowerment of the individual through the political order and his status as a citizen. In Rousseau, there is no contradiction between order and freedom, as they are simply two sides of the same coin. The practical implications of this highly abstract understanding of individual freedom in the political order are potentially worrying, as many commentators have pointed out. In fact, it can been said that Rousseau in some respects marks less a step toward modern liberalism than a full step backward toward the liberty of the ancients—that is, a step back toward a conception of social harmony that left very little room for individual spontaneity.[20]

Regardless of the nuances between all three authors, the common point is that order—political and social—is no longer the first and foremost value, as it arguably was in the ancient world. Even for Hobbes, the defender of the most absolute version of political authority, the political and social order is both chronologically and logically secondary to individuals' natural rights and freedoms. Conversely, the focus on individual rights brought by social contract theorists makes room for the possibility of individual spontaneity—in the guise of negative freedom or freedom from restraint—and even, to a degree, the possibility of revolt. The role of spontaneity as freedom of the individual is perhaps more ambiguous in Rousseau, but even he acknowledges the central primacy of individual rights. With all these thinkers, therefore, individual freedom and the social-political order are reconciled by making the political order a means to the end of a more stable, safer form of freedom, which can be enjoyed only in a polity—namely in political freedom. This freedom is bound by laws, but it is also protected and enabled by them. In other words, Hobbes, Locke, and Rousseau (as well as others not considered here) allowed Western philosophy to conceptualize the reconciliation between the collective goal of public order and fundamental individual rights and freedoms by founding the former on the latter.

None of the social contract theorists, however, provide a complete and truly satisfying answer to our problem—for two reasons. First, because their views represent an outright rejection of the holistic approaches of ancient philosophy, these thinkers have little to say about the ideal of a harmonious community, which just isn't their concern. In fact, with them the idea of an entire community flourishing drops entirely from view. Second, and quite paradoxically, despite their emphasis on the primacy of individual rights, social contract theorists end up justifying a rationalistic political order that is fairly constraining of individual freedoms. Freedom as spontaneity is either to be enjoyed in the silence of the law (in Hobbes and Locke) or exchanged for the "positive freedom" of the citizen, which is arguably much less spontaneous and, according to some commentators (for example, Isaiah Berlin), not really freedom at all. Furthermore, because entering the social contract is an act of reason and the outcome of unanimous consent, these authors can conceptualize only the possibility of an order that is the rational and planned result of individuals' conscious intentions. For an alternative model, which makes room for the possibility of order as a by-product of unconstrained spontaneities, we need to jump forward to the classical liberalism of John Stuart Mill. Mill celebrates both government by discussion (what we now call deliberative democracy) and a free market of ideas as the background for it. John Dewey and Friedrich Hayek will later develop each of these ideals in complementary ways, or so I will argue.

ORDER AND SPONTANEITY IN LATER LIBERAL AND DEMOCRATIC THOUGHT

John Stuart Mill is an indirect heir of the liberal theories of Hobbes and Locke (much less those of Rousseau). Yet, unlike those authors, as a more direct heir of Bentham's utilitarianism, he rejects the natural law tradition and particularly the idea of natural rights (whether rooted in nature or God). Mill to a degree reconciles the holistic perspective of the ancient world, for which the good of the community forms the primary standard of the good of individuals, with a liberalism of individual rights as instrumental to this common good. For Mill, the main argument for liberty as negative freedom—what I have identified here with spontaneity—is that it is essential to social utility. Finally, Mill combines this defense of individual freedom as conducive to social utility with a defense of government by discussion, which forecasts the contemporary paradigm of deliberative democracy.[21]

As far as the role of freedom in his philosophy is concerned, Mill brings us closer to a full reconciliation of order and harmony with individual spontaneity in two ways. First, he points out that the problem that preoccupied early liberal thinkers—that of the proper limits of government—does not disappear when the sovereign is the people themselves. On the contrary, there is something more subtly pernicious in the kind of tyranny that a majority can exercise on individuals compared with the obvious despotism of a king or an aristocracy. Modern representative governments are supposed to rule for the sake of all and are designed to maximize the chances that they do so (through periodic elections and other accountability mechanisms); this does not, however, mean that individual freedom is safe. From this point of view, Rousseau was wrong in thinking that the general will can never err and that it can remain sufficiently general. On the contrary, says Mill, who rehearses here an insight from Tocqueville, the problem of the democratic age is to protect individual freedoms from the people themselves, or their representatives. In this sense, Mill's definition of the harm principle—that the government is only justified in limiting individuals' freedom when our actions can harm others [22]—extends the social contract theorists' problematic to the democratic age.

The other way Mill contributes to our question is by making the case that these individual freedoms are actually instrumental to the social good. In this respect Mill entirely breaks from the social contract theorists. To make the case that individual freedoms are conducive to the greater good, Mill asks us to take a diachronic, long-term perspective on the interests of man. In *On Liberty*, he argues that allowing individuals to pursue their own definition of the good life and to express their views freely in the public sphere—no matter how improper, unconventional, or disturbing those views may be—is necessary to the discovery or rediscovery of truth. In his famous argument against the legitimacy of silencing one dissenter, Mill writes, "The peculiar evil of silencing the expression of an opinion is, that it is robbing the human race; posterity as well as the existing generation; those who dissent from the opinion, still more than those who hold it. If the opinion is right, they are deprived of the opportunity of exchanging error for truth: if wrong, they lose, what is almost as great a benefit, the clearer perception and livelier impression of truth, produced by its collision with error."[23] In other words, for Mill, freedom of expression—an outlet for spontaneity—is instrumental to the pursuit of truth. What is left unsaid in this passage, but undergirds the rest of the book, is the belief that the discovery and rediscovery of truths of all kinds are always good things overall—things that mesh with public utility and are in fact partly constitutive of social utility. To the extent that social utility includes sociopolitical order and harmony, therefore, spontaneity becomes instrumental because it tempers the risk of ossification through the rejuvenating effect of dissent, differences, even mistakes.

Importantly, Mill conceptualizes social utility as social progress, something dynamic, oriented toward the future and the discovery of new truths or the reactivation of old ones. A social order is for Mill, in essence, always first and foremost a disorderly order, or at least an evolving one. One could say that Mill advocates for individual spontaneity at the cost of some social harmony (since he celebrates eccentrics and norm breakers) for the long-term sake of a dynamic, healthy social order. Mill thus helps us think of political and social orders as no longer static but dynamic, in flux, and involving some degree of conflict, dissent, chaos, and therefore dissonance.[24]

Although Mill mostly believes that truth is best pursued by many different people in the sphere of public opinion, he also endorses representative government as government by discussion, thus suggesting that truth could also be reached through rational exchanges of arguments, not just the chaotic free flow of ideas of a liberal society. In *Considerations on Representative Government*, for example, and in spite of the title, Mill seems to have in mind the Greek ideal of public discussion among citizens (not their representatives). We thus also find in Mill the premises of what is now called deliberative democracy. To see each branch of Mill's thought come to fruition, however, we are better off turning to two very different authors. First, the American pragmatist John Dewey develops a fuller picture of democracy as a social inquiry—that is, as a quest for truth conducted not just through the deliberations of a government but those involving the entire citizenry. Second, for the idea that truth sometimes better comes about as a by-product of the interplay between freely flowing ideas, we need to turn to the economist Friedrich Hayek.

Unlike what Mill suggests (and Hayek later provides foundations for), Dewey does not think that truth can triumph simply through the disorderly clash of dissenting opinions in the market of ideas. For him truth involves consonance and, indeed, a certain degree of rational harmony. It should be pursued through political means, by discussion among an informed and self-aware public. Whereas Mill writes from the liberal point of view, Dewey is concerned with the mode of government and how this mode of government can tap the intelligence of the public, past and present.[25]

Dewey writes in the age of mass democracy and a largely absentee public—the "phantom public" denounced by his contemporary Walter Lippmann.[26] In *The Public and Its Problems*,[27] Dewey attempts, among other things, to resist an antidemocratic current according to which democracy is merely a transitory phase and government in the end is always a matter of "experts." To counter that claim, Dewey goes back to an argument in favor of democracy that he finds in Tocqueville, according to which the main strength of democracy is that it involves "a consultation and a discussion which uncovers social needs and troubles," "forces a recognition that there are common interests, even though the recognition of what they are is confused," and "brings about some clarification of what they are."[28] From this point of view, for Dewey, the main value of majority rule lies in what precedes it: "antecedent debates, modification of views to meet the opinions of minorities, and the relative satisfaction given the latter by the fact that it has had a chance and that next time it may be successful in becoming a majority."[29] Even though Dewey acknowledges the role of minorities—since "all valuable as well as new ideas begin with minorities," even "a minority of one"—he also stresses that "opportunity should be given to that idea to spread and to become the possession of the multitude."[30] In other words, even if it is true that there are only a few people with valuable insights, the value of their ideas can come to fruition only when they manage to convince the rest of their fellow citizens that they are right.[31]

The argument in Dewey can be rephrased as follows: democracy fosters discussion, which clarifies the nature of preexisting common interests and goals. In the age of mass democracy and the machine, Dewey thought we ought to return to an ideal of face-to-face deliberation because,[32] aside from the educational effect of talking to one another (also noted by Mill), truth can triumph only when everyone ultimately becomes convinced of the superior view through what is best described by Habermas as "the forceless force of the better argument."[33] In other words, whereas Mill celebrates the chaos of the market of ideas and does not focus on the active search for a consensus at any given time, Dewey invites us to look for a domain of shared views through a rational and deliberative exchange of arguments. One could say that for Dewey more than for Mill, a certain harmony of views is necessary to a well-functioning democracy. Importantly, however, and just like for Mill, the content of "truth" and the corresponding consensus are likely to change over time. Thus, while he requires moments of agreement, Dewey, as a coherent pragmatist, thinks that truth is in fact a form of opinion, mutable in nature and context dependent.[34] Social harmony is required for Dewey, but it cannot be a rigid tune.

An essential condition for the public to come to its senses and form the Great Democratic Community is, for Dewey, the communication of ideas rendered possible by art. "Democracy ... has its seer in Walt Whitman. It will have its consummation when free social inquiry is indissolubly wedded to the art of full and moving communication."[35] Like Mill in a way, Dewey advocates as a crucial condition of an effective democracy (or a "representative government," in Mill's terms) the existence of a dynamic and spontaneous civil society where the arts in particular are flourishing. Yet, where Mill thinks that true knowledge is bound to emerge out of the chaos of a "market of ideas" and thus recommends a liberal principle of laissez-faire in the sphere of ideas, Dewey accompanies his liberal recommendations with a voluntaristic program of education policies. The government, he believes, has to inculcate in its citizens if not a specific content then at least the principles of an inquiring mind and the technical skills that go with it—for example the ability to think for oneself and to defend one's views in public. One reason for the eclipse of the public diagnosed by Dewey is the incapacity of scattered individuals to find, identify, and even express themselves as a community united by a common purpose and sense of shared destiny. True democracy could be realized, Dewey reasons, if and only if people develop the skills that make them able to take advantage of their right to political participation. Whereas a certain degree of social harmony is sacrificed in Mill for the sake of truth, it is actually to be reactivated and rationally sought in Dewey. This is not to say that the two authors' views are incompatible—far from it. One could argue that Mill's market of ideas is the idea incubator for a sound and productive democratic debate among citizens. Conversely, the collective and intentional search for consensus helps acknowledge and temporarily crystallize some of the truths that bubble up from Mill's market of ideas.

Friedrich Hayek takes Mill's intuition about the epistemic properties of a free market of ideas to new heights. Hayek's views supplement Mill's in part because they offer an answer to the question: What should we seek a consensus—that is, harmony—about? What should we simply leave out of political debates? What, in other words, should be left entirely to individual freedom? The crucial part of Hayek's answer is not so much the specific reply to this question (economic transactions [36]) but rather his rediscovery of the distinction between two types of order: order that can be achieved through intentional, rational planning and order that is "spontaneous," by which he means "self-organizing" rather than natural. The first type of order he calls taxis; the other, cosmos. It is only with the first kind of order, which involves some constraints on individual spontaneity, that harmony can be intentionally sought. In self-organizing orders, however, spontaneity should be given free rein because harmony can be only its indirect by-product and not the result of a direct intervention. As a result of this rediscovery, Hayek famously offers a new definition of order that encompasses both rational and spontaneous types. An order for Hayek is thus "a state of affair in which a multiplicity of elements of various kinds are so related to each other that we may learn from our acquaintance with some spatial or temporal part of the whole to form correct expectations concerning the rest, or at least expectations which have a good chance of proving correct."[37] Hayek was writing against contemporary socialistic conceptions that intended to replace the spontaneous order of the market with the purely rationalistic one of political decision making. Such a confusion of orders—the merging of political and economic orders by having all decisions related to the latter, from production levels to prices, made by the former—is, for Hayek, both impractical and dangerous.

A rationalistic (socialist or otherwise) approach to economics is impractical in Hayek's view because it presupposes, on the part of the central agency in charge of implementing the politically designed economic order, more computing capacities than could possibly be gathered at one time in even the best minds of the best governmental teams. The impossibility of performing the calculus of the right economic order—that is, setting the price of consumer goods, the level of workers' wages, levels of production, and other economic parameters to create a well-functioning whole—is due to two things for Hayek. One is the sheer enormity of the amount of information to be processed; the other is the fact that, even assuming full information at a given time, the order is constantly evolving in such unpredictable ways that human planning simply cannot ever catch up with it. Or, to put it another way, there is no economic order outside of the actual transactions of free individuals.[38] This double complexity is why, according to Hayek, it is better to let that infinitely complex calculus be performed at every instant by the unthinking mechanism of the market.[39]

Such a man-made order is unable to deal with the infinite complexities of human interactions and thus risks impoverishing them by too much regulation. But it is also easily threatened by manifestations of dissent, which pave the way toward more and more control and toward the transformation of the government into an authoritarian and potentially totalitarian organization. In that sense, the socialistic order Hayek was fighting against shares traits with the ancient order that characterizes, for example, Plato's aspirations for a just republic (except for the class distinction). Socialism seeks to implement an egalitarian society in highly rigid, controlled ways that ultimately sacrifice individual spontaneity to the collective goal of social harmony.

In light of the complexity and emerging nature of the economic order, and the dangers of trying to design and control it, Hayek thought it best to keep the interactions of individuals in the economic sphere as free from rationalistic intervention as possible. Of course, the extent of the "possible" here is bound to remain a matter of controversy. One may agree with Hayek's view of the market as a spontaneous order (as I think one should) and yet disagree with the extent to which it could be corrected by rational decisions. The debate crucially depends on empirical claims about the imperfections of actual markets and how rational intervention—political regulations—can help fix them. But the original insight remains powerful and I believe true: In some domains of human life, including economic interactions, order and harmony are best attained as the result of horizontal and free-flowing individual spontaneities rather than a rational order imposed from the top down, even one that is the result of democratic and open-ended deliberations.

What Hayek brings to the table, therefore, is a conceptualization of order as dynamic, self-organizing, and constantly self-rearranging. It is a horizontal concept of order, by contrast with the hierarchical, vertical concept of the ancients and even of the social contract theorists. It is also a spontaneously emerging order, in contrast to the rational one of deliberative democrats.

CONCLUSION

We started by noting the obvious tension between the concepts of order and harmony on the one hand and those of spontaneity and revolt on the other. This brief journey through moments of Western thought shows that the tension can be partially resolved, although usually by leaving one term outside of the equation. We saw that the social contract theory reconciled political order and spontaneity, by making political order an instrument for the preservation of some degree of negative freedom. The theory also, in its Lockean version at least, made room for the possibility of revolt and political change. The reality of social contract orders, however, was to minimize spontaneity by rendering it possible only in the silence of the law. As to revolt, it was the necessary counterpart of a political order so rigid as not to allow for easy change. With the rise of representative governments, the political order becomes more flexible and responsive to social changes. The problematic of John Stuart Mill becomes that of integrating individual spontaneity with the social order by showing how it is instrumental to human progress. The notion of a rigid, hierarchical political and social order gives way to a more flexible, dynamic kind of order, where harmony is no longer the primary value. Society becomes a space where individuals should be encouraged to express themselves for the benefit of all, even when they clash against established norms. Within that liberal order, Dewey and Hayek help us refine that new concept of order. Dewey reintroduces the possibility of harmony with the search for a deliberative consensus that feeds off the collective search for truth. Hayek suggests that in some domains of human affairs, particularly economics, society is better off letting individual spontaneities play out and produce their own unpredictable harmony. Among these more recent thinkers, the concept of revolt is largely absent, because in their more organic view of social order, there should be no need for violent disruptions. Revolt is civilized by Mill into demonstration, protests, civil disobedience, petitions, or simply a minority vote. In Dewey it can be more efficiently expressed as disagreement in the collective discussion. In Hayek's view of the economic order, revolt has no place at all.

Each of these three thinkers, it can be argued, has conceptualized the relationship between social or collective order and individual freedom and spontaneity with respect to specific tenets of modern Western ideology: liberalism, democracy, and the free market. All three authors give us a story about how to reconcile the tension between the stability and predictability wanted in an orderly whole and the disruption and unpredictability inevitably brought by individual spontaneities. What is interesting is that in all three models—the Millian order of a liberal society, the Deweyan order of a democratic society, and the Hayekian order of a free market—individual spontaneity is not meant to be tamed, reduced, or formatted, except under specific circumstances. It is instead encouraged, cultivated, or more neutrally harnessed. What gets sacrificed, perhaps, is a certain type of order and the static harmony that goes with it. Order, for those thinkers, comes as a flux of contradictions and disagreements that resolve just in time to be reborn in a different form. From an individual perspective, this may give an impression of disorganization and chaos, yet it is an order arguably more stable and resilient over the long term than the static and surprisingly brittle orders of the ancient world or even the rationalistic order of the social contracts. It takes perspective and time to learn to appreciate the unplanned regularities in the apparent chaos of liberal societies, democratic public opinion, and the free market—a bit like learning to appreciate free jazz's violation of harmony in favor of a new kind of free and spontaneous musical order.

CHAPTER 7

Japan: The Romanticist Revolt against the Empire

Alexander Dolin

For more than two and half centuries, for the Japanese nation, isolated from the rest of the world, the rule of law meant the enforcement of peace through adherence to a rigid social order. That kind of social harmony was an imposed and extremely hierarchical form of harmony that certainly did not suit the desires of all members of society. In this paper I will consider the more spontaneous and revolutionary ideas from the West absorbed by the Japanese Romanticists and developed further in their confrontation with the imperial order.

The reign of the Tokugawa shoguns was based on fear and duty imposed from the top by the Confucian-oriented government. Every step of the loyal subjects was monitored by the secret police. Spying and reporting on neighbors were encouraged. Any deviation from the prescribed rules and regulations could be punished by imprisonment, torture, or death. Writers and painters who dared to mock the establishment in their works were severely persecuted. Any revolt against the system was doomed and could end only in death. This very topic—the tragic fate of lovers who tried to break conventions—became a favorite theme of Edo novelists and playwrights. In the early eighteenth century, the protest of forty-seven *ronin* against an unfair verdict to their lord issued by the shogun, which took the form of a bloody vendetta, ended in a mass suicide that made them national heroes. Not a shadow of freedom could be allowed in the samurai world of the shoguns. It took many years before freedom of thought and true democracy were imported to Japan from the West. Then new revolutionary forces emerged in the country, which was awakening from a long slumber.

The Meiji Restoration (1868) not only opened Japanese ports for Western ships but also initiated an influx of "Western knowledge," which resulted in comprehensive reforms of the administrative system, social order, and economy. From the very beginning, the reforms were aimed at the construction of an Eastern military superpower. However, the attitude of new Japanese intellectuals toward Western culture and art was quite controversial. A large part of the young literati gravitated toward the West and advocated a policy of integration with the world community. Meanwhile, their opponents supported the government in the endeavor to create a model totalitarian empire on the basis of revitalized traditional Confucian and Shinto values. In the late 1880s, which became a period of fast transition to a modern industrial society, a fierce debate was launched between adherents of a cosmopolitan ideology of Westernization—writing in the journal *Kokumin no tomo* (*The Friend of the People*), founded by the progressive philosopher Tokutomi Soho—and their opponents, nationalists from the Seikyosha (True Teaching) Society, united around the magazine *Nihonjin* (*The Japanese*).

Members of the Seikyosha group vociferously opposed servility toward the West and called for the "preservation of national identity" (*kokusui hozon*) within the framework of the existing social order in Meiji Japan. In his book *The Truth, Kindness, and Beauty of the Japanese* (*Shinzenbi nihonjin*), Miyake Setsurei, a talented propagandist of rising nationalism, claimed the superiority of the Japanese over other nations and in particular over the white race, thereby assuming the priority of Japanese culture. This racial theory had further development in the works of another member of the Seikyosha group, Shiga Shigetaka, who grounded his concept of the chosen nation of Yamato (the classical name for Japan) in geographic, climatic, environmental, and historical factors. Due to the efforts of members of Seikyosha, the idea of "people's nationalism" (*kokuminshugi*), a close relative of the concept of state nationalism (*kokuseishugi* or *nihonshugi*), gained a certain popularity. Nevertheless, Westernization had already taken firm root in Japanese soil and would not easily yield ground to conservative views.

Religious reformation also played a crucial role in the transformation of the ideological foundations of Japanese society. After a ban on Christianity was lifted in 1873, the activities of English, French, and Russian missionaries resulted in the proliferation of Christian dogma among the new Japanese intelligentsia. In the 1880s, several Christian schools, colleges, and universities, and 130 churches, were founded in Japan. The Bible was translated into Japanese (for the first time since the sixteenth century), and its most lyrical part, the Psalms, was widely exploited as a source of church hymns. The new faith was promoted throughout the country by numerous Christian organizations, and many of the most active intellectuals came under the strong influence of Christian ideology, which was for them the embodiment of Western cultural values. However, the pro-Western views of the newly converted Christians and young sympathizers faced active opposition from a conservative majority that adopted a positive attitude toward the mainstream nationalist doctrine of *kokutai* ("national polity"—that is, the state as an integrated body). The biblical texts were opposed by the canons of state Shinto, Confucian dogma, and the severe regulations of the Imperial Edict on Education (1890). The aim of the government was to work out a new ethical system that would mobilize citizens to serve their country and their sovereign. Thus the new intellectuals of the Meiji period had to define their role against a backdrop of modernization in which the cultures of East and West were destined to act like yin and yang, polar forces continually complementing and struggling against each other.

IMPERIAL AMBITIONS AND ROMANTICIST IDEALS

Accordingly, all aspects of culture—science, ethics, philosophy—are endowed with values only as far as they promote the realization of these values in a group—but not individually, not inside one's own mind. That is why the so-called spiritual revolution—the revolution in public consciousness during the third decade of the Meiji period—influenced, first of all, views on the role of personality in public life and the creation of artistic values. It marked the transition from the conventional corporate mind to the individual mind, from medieval "school and guild" art and literature to a modern individual artistic mentality. One of the major theorists of Japanese Romanticism, Togawa Shukotsu, writes: "So a person is always in search of his 'ego,' but one's personality is a part of the universe, and the universe is a part of one's personality. Actually, one's personality cannot be comprehended absolutely, and that is why, in unveiling the secrets of nature or studying history, a person perceives only himself. Isn't it better to turn from studying the surrounding world to perceiving one's inner world?"[1]

FREEDOM OF THOUGHT AND SOCIAL DUTIES

The ideas of Togawa Shukotsu clearly reveal the influence of European subjective idealism. But while Shukotsu accepts the importance of personality in general, along with other Japanese Romanticists he tends to reject the concept of elite art, the theory of the "genius" as an exceptional personality. The Romanticists were eager to apply to their own country the progressive elements of concepts put forth by their Western forerunners, but in a country bound by Confucian morals, burdened by patrimonial and class hierarchy, the very declaration of personal freedom meant an open attack on public consciousness and on the conservative ideology of "Japanism" promoted by the government. Thus the Japanese Romanticists never associated personal freedom with freedom from society. On the contrary, they treated a personality as an inseparable part of society and considered instead revolt against spiritual and physical slavery as their primary goal in constructing new social ethics.

"The truth is that a human being should defend freedom," argues the leader of the Romanticist school Kitamura Tokoku in his "Ideals of Common People in the Tokugawa Period." "If we study more attentively the course of history up till now, how many examples we will see illustrating that freedom was gained at the cost of blood, endless suffering, and torment!"[2] The struggle for freedom by means of "the brush and the sword" is a major topic in Tokoku's works, although it is often accompanied by the gloomy premonition of an inevitable defeat. While any attempt at democratization of the social order in those years was inevitably doomed to failure, the democratization and emancipation of the individual mind in the milieu of the Meiji intellectuals was quite successful due largely to the influence of the Romanticists' works.

In his article "The Perversity of Blind Acceptance and Groundless Negation" ("Ganshu mohai no hei"), Tokoku reveals a considerable understanding of the contradiction between the traditional Japanese "pagan" model of the uniform world and the dual philosophy of Christian idealism, clearly expressing his sympathy with the latter. As a matter of fact, the problem expressed by Tokoku—that is, the issue of the evolution of ethical priorities in society—was elaborated by many theorists of European Romanticism and originated with Schiller's article "On Naive and Sentimental Poetry," in which by "naive" poetry he meant primarily the literature of pagan antiquity and by "sentimental" poetry he meant that of emerging Romanticism.[3] This problem was repeatedly studied by the scholars of Western Romanticism,[4] but with reference to Japan, their conclusions definitely require some amendments.

First, Japanese Romanticism emerged later than similar schools in the countries of Europe and America. Secondly, Christianity was not capable of dominating and changing completely the worldview of writers who lived in a country with a different religion (or, to be more precise, different religions). Thirdly, due to a shift in chronological borders and the accelerated pace of the formation of new schools, Japanese Romanticism could at the same time support many rationalistic ideas of the Enlightenment, unlike the case in Europe, where Romanticism represented in a certain sense rejection of the Enlightenment.[5] And finally, the traditions of Japanese classical literature and folklore, due to their specific character, differed considerably in spirit from the European classical heritage. All this motivated an alternative attitude among Japanese Romanticists toward the problem of "naive" versus "sentimental" poetry.

As we know, in Europe in the eighteenth century, the theory of "natural rights" gained great currency. The idealization of a "natural" condition of mankind in a patriarchal, classless society, such as Rousseau's call for a "return to nature," influenced all the bright minds of the time.

Schiller, however, rejected the idealization of a "natural" society. He criticised Rousseau and his followers for their attempt to lead mankind back to a primitive condition lacking high ideals and true cultural values. Schiller warned young Romanticists against excessive worship of a supposedly "harmonious" antiquity. Later the Romanticists, referring to the issue enunciated by Schiller, expressed regret concerning the tragic split in modern culture that resulted in the introduction of pessimism into Romanticist writings.

A. W. Schlegel formulated the differences between "naive" and "sentimental" art: "The Greeks assume that human nature finds satisfaction in itself. They did not feel any uneasiness and never aspired to any other perfection, except the one they could gain themselves.... With the advent of Christian views everything changed: the contemplation of the infinite destroyed the definite; life turned into the night world of shades, and only the next world revealed the dawn of true existence."[6]

OBJECTIVES OF THE ROMANTICIST REVOLT

All classical German idealism is based on the conviction that the ideal in real life is unattainable, but it acknowledges poetic revolt and escape in the realm of Beauty as the only way to face the brutal reality of the philistine world order. This doctrine defines in general the concept of the individual in Romanticism, as well as the attitude of many Romanticists to art and to creativity in general.

The Japanese Romanticists, despite their fascination with Schopenhauer and Hartmann, nevertheless criticised the extreme idealism of their views. No wonder Shukotsu urged his compatriots to protect Japanese literature from the influence of Western decadence. Japanese literati regarded their "revolt" not as a desperate struggle against laws and regulations but mostly as an effective means to implement and promote spiritual revolution for the benefit of the nation. Of course, inability to achieve these ideals often turned into tragedy, but it's no doubt that the best masterpieces of Japanese Romanticism are charged with a positive "energy of creation." As active players in a young, developing society, the Romanticists claimed that the basis of life and poetry was enthusiasm or zeal and that no creativity was possible without it. In his essay "Enthusiasm" ("Netsui"), Tokoku asserts: "If we withdraw such a basic element as 'enthusiasm' from human life, the poet in his activity will be not able to bear the burden of glory. If humanity did not possess such a category as 'enthusiasm,' mankind would not have any history, and men would resemble four-legged animals."[7]

It is by assuming this perspective on enthusiasm and on the necessity of fighting for one's ideals that Tokoku attempts to overcome the narrow-mindedness of a more pessimistic Western Romanticism.

An explicit desire for social progress drives the Japanese Romanticists away from the idealization of ancient and medieval culture. (The role of "antiquity" in the theory of Japanese Romanticism is assigned predominantly to the culture of the Heian and Kamakura periods.) The stability of medieval mentality, based on the alleged harmony of social conventions and severe canonic restrictions, was alien to the Romanticist world. Confucian dogmas of the samurai code of honor, imposing Buddhist idealism and Shinto animism, were condemned by Romanticist writers as the pillars of the conservative old-fashioned social order. They rejected point blank the elaborate harmony of centuries-long social relations based on rigid hierarchy and especially formal family relations lacking true love. They were attracted first and foremost to new, progressive ideas. Their artistic creativity was considerably influenced by Rousseau's views, but at the same time they could not but admire the muse of the nonconformist Zen poets of the Edo period who had achieved absolute freedom in their spiritual experience. As Shukotsu writes in his essay "Reflecting on the Characters of the Haikai Poets" ("Haijin no seiko o omou"):

> They wander in the heavens yet feel no restraints imposed by moral dogmas. They breathe the air of freedom, avoiding all moral restrictions [the rigid prescripts of Confucian ethics]. They do not care if it will bring them disgrace. They reside in a different world, described by an ancient poet:

The peach blossoms fall
And disappear
Into a different world
Where no human can be seen.

The definition of their world is in these words: "The happiness in which the soul bathes is the feeling that overwhelms the precision of thought." Those who mock them may rot in misery. Who will take over the accomplishments of those poets in the Meiji era?[8]

According to Shukotsu, the concept of "natural inspiration" is embodied in a person nourished by the spirit of free creativity, who lives an intense "inner life." Shukotsu's ideal is that of an independent intellectual, a *bunjin* of a new type, one who has mastered Rousseau's ideas, placing spiritual freedom above the laws and regulations of society. His ideal is not only a Zen *haikai* poet but also Li Po, who was considered "an immortal, expelled from heaven," as well as Byron and many other "spiritually emancipated" poets. According to Shukotsu, writers who continue the traditions of the poets of *haikai* verse in the Meiji period must combine in a harmonious way their creative individuality and the quest for the Romanticist ideal with the spiritual freedom inherited from their ancestors, who were content with the minor joys of life.

At first sight, Shukotsu's call to follow the lead of the ancient sages, submitting only to the needs of the heart, reminds us also of the slogans of European Romanticists, who called on the poet to step away from "the crowd" and the "common" people. However, the intention of Li Po and Japanese Zen poets to cultivate their inner lives and "merge with nature" is no more than a poetic declaration, behind which there is a sensation of ties with the people, the call of "blood and soil." For example, it is difficult to find in the history of Japanese literature a writer closer to the common reader than Matsuo Basho, who lived in a lonely "banana-tree hermitage" and wandered the mountain paths far from the cities. In fact, the genre of *haikai* emerged in the milieu of "the third class" as a late medieval reaction to the development of the aristocratic poetry of *tanka*.⁹ Thus the essay by Shukotsu points more at the ideal of a democratic art than at the elite concept of creative activity so popular in the West since the period of Romanticism.

Kitamura Tokoku, in his article "On the Inner Life," more accurately articulates the concept of "ideological art," opposing both the pragmatic approach to literature found in the teachings of Yamaji Aizan and a complete withdrawal from reality. Tokoku confronts his poetic ego to any imposed artificial conventions and restrictions, asserting that the basic mission of art is to elevate the human spirit, appease human suffering, and open the path to spiritual purification. He claims that the writer's role is to express truth and justice: "A poet's goal is to convey in his words the inner mind and internal experience, modified in a creative way. The highest mission of the poet, who is also a philosopher, lies in the opportunity to tell the readers about his own inner life"¹⁰

Tokoku urges writers to turn from describing sketchy characters and situations to solving important and complicated social, political, and philosophical problems by means of art, where the author projects his "internal life" on the phenomena of the surrounding world. At the same time, Tokoku defines the concept of "prose fiction" (*bungei*) and claims that its function differs essentially from philosophy, on the one hand, and from entertainment, on the other. The noble principles proclaimed by such literature have to be vested in an adequate form: "It is neither necessary nor possible for literature to turn directly, like religion or philosophy, to vital problems. Literature is a phenomenon that unites ideology with art. If a work has ideas but lacks art, it is not fiction; if a work has art, but has no ideas, it can also not be fiction."¹¹

It was a novel definition of the mission of literature, in particular fiction. He intended to elevate literature from the status of "pulp fiction" (the destiny of the *gesaku* stories and novels in the Edo period) to the status of a manual of life. Tokoku also designated an absolutely new role for poetry. Before the Meiji Restoration, Japanese literature was regulated by a complex of antinomies, such as *graceful/vulgar* (*ga/zoku*), *high/low* (*ki/san*), and so on, and thus the introduction of a generally democratic Western poetics as a model was apt to cause a shock. The new poetics did not manage to reconcile long poems (*shi*) with traditional short *uta*. Even on a new stage, Japanese *shi*, which was already transformed in *shintaishi* and had completely lost its resemblance to *kanshi* in Chinese, would remain a strictly professional genre, meant for experts and their fans. Only in due course, thanks largely to the efforts of Kitamura Tokoku,

Shimazaki Toson, and Doi Bansui, was *shintaishi* turned into a truly national poetry, *shika*, which "emerged from a complicated process of synthesis of the Chinese poetic tradition and European verse."[12]

Tokoku's understanding of the categories of form and content is rather pragmatic. While some European Romanticists are called the forerunners of "pure art," Tokoku and his followers did not at all cherish such an ideal.

Shelley, arguing for the priority of poetry over all the other kinds of art, claimed that "poets are the unacknowledged rulers of the world."[13] His words are in tune with the views of Tokoku concerning the role of literature in transforming the spiritual world of an individual in the new age. In the article "The People and Ideology" ("Kokumin to shiso"), Tokoku urged Japanese writers to promote progress and to reveal the national identity of people who inhabit the country, since it is they who possess the specific spiritual values that unite and create the nation. Tokoku sees in the people the source of creative power, the creative energy that would feed the new culture born of the fusion of East and West.

However, Tokoku's goal of constructing a new doctrine of national art remained incomplete for two reasons. First, although he believed he was acting on behalf of the whole nation, he actually expressed the interests of only advanced intellectuals. While he reflected on the misery of the common people in the Edo period, he failed to evaluate the real situation of the masses after the Meiji Restoration, the period that saw the beginning of modernization and the accelerated growth of capitalism. Secondly, while calling for a native approach in literature, Tokoku at the same time sharply criticized the Movement for the Unification of Literary and Colloquial Language (*gembun itchi*) initiated by Futabatei Shimei. He in fact set the pattern for other Romanticists of writing poems in the traditional *bungo* style, which was difficult for the common reader to understand.

Tokoku's fascination with the history of the Japanese people and his acquaintance with the works of the great Russian writers of the nineteenth century, such as Tolstoy, Dostoevsky, and Turgenev, induced the Japanese author to look, at least to some extent, on the problem of "the individual and society" from a materialistic point of view. Under their influence, he introduced in his Romantic writings a strong element of materialistic criticism. In his article "The Murderer's Crime in *Crime and Punishment*" ("*Tsumi to batsu* no satsujin tsumi"), Tokoku argues that human characters depend on social conditions and not on innate immorality or morality. In his critical analysis of the novel *The Torments of*

Hell (*Abura jigoku*), written by his contemporary Saito Ryoku, a member of the Kenyusha group, Tokoku speaks against the false conventions of contemporary pseudo-classicism, calling for literature to be brought closer to public life. With this slogan he participated in the Movement for Freedom and Civil Rights, trying to build a democratic opposition to the conservative Meiji government. Yet such ideas in Tokoku's works are more often than not contained within naive Romantic forms. They have only a faint resemblance to the principles of realism developed by the novelist Futabatei Shimei on the basis of his knowledge of the Russian classics.

INSPIRATION IN FAITH

Despite his infatuation with Russian classics, Tokoku definitely prefers the Romanticist method of depicting reality. In his essay "Passion" ("Jonetsu"), he argues that realism, which is supposed to reflect everything faithfully and objectively, lacks real passion and inspiration. He claims that a true Romanticist, like any other great master, can really be inspired only by a deep and sincere belief: "A great poet always has his own belief, his faith, his own religion, his own understanding of the divine. In Homer we feel the spirit of the ancient Greek gods, in Shakespeare we feel the faith of medieval England; Saigyo has his own religion, and so has Basho."[14]

A Christian himself, Tokoku did not insist on Christian monotheism when it came to art. He admits the right of any artist to have his own faith, one that represents a certain theism, even if it's not necessarily a religion in the proper sense of the word but simply a form of spirituality, an embodiment of the unconscious. Influenced by Hartmann's *The Philosophy of the Unconscious*, Tokoku assumes the primacy of the irrational over the rational. While in the field of ideology, the Japanese Romanticists managed to overcome the gloomy pessimism and passive worldview of Hartmann, in the area of creative technique, the "philosophy of the unconscious" provided them with a model, an outline of the intuitive approach to reality. In fact, that was the focus of the Romanticist challenge to the traditional ideology of Confucian order and in a certain sense also to the Buddhist quest for harmony beyond the mundane passions. Even such a progressive Romanticist as Tokoku, whose goal was to fight for national ideals, follows the ideas of classical German idealism in acknowledging the unconscious as the only means of achieving his goals.

At the same time, it is important to remember that intuition also lies at the core of traditional Buddhist metaphysics and that it thus forms part of the essence of the centuries-old Japanese literary tradition. If we compare in particular Zen intuition to the irrationalism of Western philosophers, we will be able to comprehend more distinctly a problem that the Japanese Romanticists and their successors had to face when choosing their creative techniques.

According to the principles of Zen aesthetics, any creative activity has to be impulsive, spontaneous, influenced by intuition, and directed at establishing contact with an external world by involving a counterpart (a listener, spectator, or reader) in an interactive creative process. Absolute freedom and naturalness are the necessary prerequisites of the deepest self-realization and self-expression.[15] The indispensable condition that makes it possible to create a truly valuable artifact or poem is the earlier accumulated potential that a master has acquired in the course of lengthy practice. An ingenious, impromptu masterpiece is unattainable for a layman or a novice in art, just as a neophyte in Zen cannot reach satori. The priority of feeling over reason, of intuition over rational knowledge, provides Zen aesthetics with the same features of subjective idealism that attracted the Romanticists to the theories of Schopenhauer and Hartmann.

The Russian scholar F. de la Bart defines the basic concept of creative activity from the perspective of European Romanticism in the following way: "Only the mystically inspired feeling of a poet is able to conceive the perfect ideal.... It is a feeling that discloses the meaning of reality, which in itself is dead"[16] This definition can also be applied to the ideas of the Zen masters. The difference lies only in the final goal of the act of creation. A Zen master's goal is in his creative activity per se; he does not separate aesthetics from ethics. The act of creation becomes, on the one hand, a way of self-expression and self-improvement of the artist's personality and, on the other hand, a way of being united with the world, with nature, and with others. An artist is not eager to escape from reality, but he also does not seek to solve the world's problems or influence his reader intentionally. The poetry of allusion that emerged on the basis of the intuitive perception of the world appeals, in its turn, to the intuition of the counterpart (the reader or spectator), inviting him to participate in the discovery of the "the pathos of things" and the hidden, mystical sense of life.

Of course, Western Romanticists perceived the goals of their creativity differently. Wordsworth, for example, rejecting the concept of Romantic revolt against society, believed that the role of the artist is to preach moral values, teach about life, purify the spiritual world of a man with beauty—

just as religion does—and bring truth to the hearts of readers. Edgar Allan Poe, on the contrary, stood for "pure art" and rejected any moral imperative in poetry. We will not try to specify to what extent the work of either of the poets objectively corresponds to their ideas. The important thing is that all Western Romanticists are united by their aspiration for the ideal, by their ardent desire to taste the mysteries of the spiritual life. In Japan, before the Meiji Restoration, "a writer was more interested in the supreme laws of reality than in a person himself or his inner world. The person became a symbol and stopped being a real person living on the earth."[17] The cultivation of the idea of a perishable world, the absence of tragic conflict (as noted by Tokoku), the rejection of passions in favor of simple emotions and mundane problems—these are the features of traditional Japanese literature that curbed its intensity and fettered its psychological attitude.

There was no place in the traditional system of spiritual values for an active transformation of the world. Generally speaking, the core principle of Zen aesthetics that was rooted in Japanese art deeper than any other aspect of Buddhism can be considered a doctrine of the golden mean, a way for humans to achieve harmony with the world. This approach was attractive to many at various historical junctures and stages of social development. But by the end of the nineteenth century, the Buddhist ideal of non-action (Chinese *wuwei*; Japanese *mui*) had been exhausted, and this fact was reflected not only in the works of the Japanese Romanticists but (later) in Chinese Romanticism as well.[18]

The Romanticists, for the first time in the history of Japanese art, aimed at perceiving the human being as an individual, exposing his psyche, his inner world, his thoughts and sentiments. That approach to human ego was a revolution in itself. They turned to the idealist doctrines of the West and realized that the old methods based on an intuitive, sensual perception of the world were also quite admissible and helpful. However, the artistic techniques, methods, and conventions of the past were not sufficient or acceptable for a new art. This revelation was crucial for modern Japanese literature in terms of finding the way to further development. Even if the Romanticists had not managed to create outstanding art themselves (that is, poetry of undeniable beauty), their achievement, from the point of view of literary history, would nonetheless have been of great significance.

CREATIVE SYNTHESIS AGAINST THE CANONIC REGULATIONS

The Romanticists rightly observed that the transition from "non-action" to action and the creative activity of an artist implied considerable reduction of the suggestive element in verse. Regardless of whether a poet wishes to transform the world or to escape into the sphere of "pure art," he needs the relevant effective poetic techniques. One can express "eternity in a flower cup" by one stroke, but conveying the doubts and disappointments of the restless soul of a poet seemed possible by means of only abundant imagination and skillful description.

If we treat the creative act not as a means of perceiving the ephemeral charm of an object but as the way for an artist to achieve self-expression through the comprehension of the beauty of the world, then poetry, according to Shelley, can prevail over the curse subordinating us to casual impressions of reality. Even when trying to merge with nature, the Romanticists do not focus on isolated, particular objects but are eager to generalize and perceive the beauty of nature at large. This approach was applied to not only perceptions of nature but all spheres of Romanticist poetry. For the Western poet as well as for the reader, especially in the period of Romanticism, elevated poetry always represents a spiritual quest, a kind of religious rite. But in Japan, classical poetry was always only one way to achieve a natural harmony and peace with the world. We should not be surprised by the comparisons of Western poetry with a palace or a temple, by attempts to draw a parallel between poetry and architecture or painting. In fact, the same situation can be observed in the attitude of the Japanese toward traditional poetry, which has always existed in a harmonious combination with painting, calligraphy, ikebana, and so on. However, the general aesthetic principles that shape the foundations of art forms in the West and in Japan are different. The magnificent massiveness of medieval European architecture is incompatible with the transparent lightness of structures made of wood and paper. It would be just as inappropriate to compare the paintings of Poussin with the landscapes of Sesshu. The same holds true for poetry.

The Romanticists, reconsidering aesthetic principles in general and penetrating into the nature of the creative activity of their European predecessors, assumed from the very beginning that poetic minimalism and the consistent maintenance of self-restrictions in matters of technique and form had become obsolete. The new age needed new songs, not restricted to any old codified canonic rules. After all, the Western poets treated that kind of laconism critically. According to Keats, poetry should first of all impress by its beautiful excessiveness, not by its singularity. Secondly, in its beauty there must not be any reticence that takes a reader's breath away but does not leave him contented. The images should rise and move before him naturally, bringing light and then dying away in strict solemnity and magnificence, like the sun, leaving him in a wonderful twilight.[19] The allure of such a prospect so captivated the imagination of young poets in Japan that they dared to sacrifice the rigid refinement of the old forms for the sake of "exquisite excessiveness."

Romanticist *shintaishi* poetry emerged as a natural result of the search for descriptive forms capable of containing the abundant feelings and revolutionary ideas of a person of the new age. This search partly succeeded, but the lack of a solid foundation, the weakness of the tradition, inevitably led to the unsteadiness of the acquired forms. Having abandoned the well-balanced harmony of the old art, the Romanticists could not carry out their ideal to the end and achieve real perfection in their new undertakings. Like the giant Antaeus, who remained invincible as long as he was touching his mother, the earth, but lost his power when Hercules lifted him up in the air, a poet is strong and almighty only until he leaves the ground of reality; he becomes powerless when he starts to float in a blue fog.

The success of the Romanticists' best works can be attributed to a natural synthesis of Japanese and Western civilization—that is, to precisely what the champions of the "spiritual revolution" had hoped to achieve. No wonder that this was their primary goal after Japan's many centuries of total isolation from the outer world. Intercultural exchange became the motto of the new Japanese intellectuals in their revolt against the traditional "islanders' mentality." A similar "aesthetic relativism," opposed to narrow-minded nationalist prejudices, was typical of the Western Romanticists in the early nineteenth century.

One of the pioneers of Romanticism, Kunikido Doppo, wrote in the preface to his first poetic collection: "Received as heritage, we feel in our veins the blood of ancient sages and in the depth of our soul the Oriental sentiments are burning. According to the records, we were converted to Christianity, but in our tiny hearts we feel the struggle of the feelings and thoughts inherited and acquired in the process of education and which were transmitted to us from both the East and the West. When you want to imagine a rainbow, you can read the elevated poetry of Wordsworth; when you hear the evening bell, you remember the sad lines of Saigyo."[20]

The interest of writers in the past, as was proven by the experience of European literature, was one of the characteristic features of Romanticism. Romanticist writers in the West, having broken the bonds of classicism, rediscovered for themselves the works of antiquity and the Renaissance. In the meantime, they tended to go across borders, seeking inspiration in Oriental myths, legends, and lyrical masterpieces. Likewise, Japanese writers inevitably came to the study of the culture of the past in their quest for the Romantic ideal. Only an indigenous aesthetic heritage, not one imported from abroad, could become the true foundation of a new Japanese literature. Christian humanism was just a shield that protected the inner world of the Romanticists in their struggles. The world concealed under this shield was rich with emotions and bold aspirations, ready to embrace all the spiritual treasures of the earth. The growth of individual consciousness in the period of "spiritual revolution" defined the negative attitude of the Romanticists toward the problem of the poetic canon. Most scholars believe that traditions before the Meiji Restoration in general did not reflect the individual mind appropriate for the new age.

Romanticists appreciated most those authors they considered forerunners, whose works were marked by a conceptual boldness or distinct individuality. They singled out authors who did not accept conventional restrictions, such as the Zen master Ikkyu (fifteenth century), and those who became founders of new schools (for example, Basho). This selective approach was also applied to classical Chinese poets, especially to poets of the Tang era: Li Po, Du Fu, Bo Juyi, Wang Wei, and Meng Haoran. Here again, as with the attitude of the Romanticists toward the heritage of the European Renaissance, we find poets who lived in periods of cultural revival and who dedicated their careers to the promotion of humanist ideas.

Thus, if we assume that the creative writings of the Romanticists were fed by two sources—the culture of the East and the culture of the West—we should include in the first domain, along with the works of Japanese writers, the heritage of Chinese classical literature and philosophy. However, Romanticists in Japan did not overestimate the role of this heritage. They did not overtly oppose the classics to modernity or try to establish the "cult of the past" as their Western counterparts, the Sentimentalists and some Romanticists, did in Europe. (One should remember Macpherson with his *Ossian* in England and Uhland with his "medieval" ballads in Germany.) For Japanese Romanticists, it was important to master equally the spiritual legacy of the West, the native literary heritage, and the Chinese classics. All three components would contribute to expansion of the horizons of the new literature, forming a strong background in its struggle against canonic restrictions and regulations.

CHALLENGING TRADITIONAL MORALS

The discourse on the topic of love and relations between men and women in Meiji Japan was probably even more vital for the formation of the new mentality than the same theme had been in the literature of European Romanticism, since it was the issue of love, as one of the main problems of aesthetics, that met the strongest resistance in a society not yet freed from feudal attitudes. Choosing love as an object of discussion, and then trying to investigate its nature and define its role in the life of the individual, the Romanticists thereby challenged the Confucian morals and conventions of the past that still dominated Meiji Japan.

To be sure, love as a topic of literature was never under a ban. One might recollect love poems from the *Manyoshu*, the *Kokinshu*, and numerous other anthologies or the *Ise Monogatari*, as well as the plays of Chikamatsu and many other masterpieces of the Japanese classics. Still, a real Romantic cult of love was basically impossible in premodern Japan, where Confucian values always prevailed over individual passions and emotions.

Romanticists first of all rejected the Confucian formula of female virtue: obedience to the father, obedience to the husband, and obedience to the eldest son. From beings of a subordinate nature—either as humble housewives or as objects of momentary pleasure—women were turned into vessels of supreme spirituality, at once objects of worship and sources of poetic inspiration. In other words, for the first time, the cult of the Fair Lady, la Belle Dame—to a large extent borrowed from the West—emerged in Japan. Idealists such as Tokoku saw in love both a universal force guiding the spiritual life of human beings and an archetype of the differences and similarities between the individual mind in the East and the individual mind in the West (with preference given to the latter). Tokoku wrote about love often, not only in letters to his wife-to-be, Ishizaka Minako, but also in a number of philosophical essays, including "The Poet-Pessimist and Women" ("Ensei shika to josei"), "On the Inner Life" ("Naibu seimei ron"), "Contemplation of Another World" ("Takai ni taisuru kannen"), and "Ideals of Common People in the Tokugawa Period" ("Tokugawa jidai no heiminteki riso").

A Romanticist to the marrow, Tokoku perceived love as a certain ideal existing in the material world but in no way belonging to it, since it was a manifestation of supreme spiritual forces: "Can a man, whether young or mature, whose sincere convictions in this transient world lead him to pessimism, and who is unable to overcome that pessimism either by faith or by life experience, find something that is not just a transient illusion? Yes, there is such a soothing element, one which appears truthful, permanent, and immortal: it is love."[21]

From his standpoint as the champion of Platonic love, Tokoku criticized the fiction writing of the Tokugawa era, which promoted at once hedonistic views and pragmatic values. He was particularly critical of the aesthetic principle known as *iki* or *sui* (literally "elegance" or "stylishness"), which became the foundation of the *ukiyo* (transient world) culture of the Edo period and which defined the approach of bourgeois commoners to love and the major problems of life in general. According to the definition of Hisamatsu Senichi, the ground for development of the *iki* principle was created by the drive for money and sexual pleasures in the "licensed quarters." Eroticism or lust (*koshoku*), in other words, helped to feed the drive for *iki*, and townspeople who considered *iki* their life philosophy were eager first of all to become rich and enjoy life in this impermanent world. Traditional aesthetic categories such as *sabi* (the sad charm of existence) and *wabi* (simpleness; rusticity) were more likely to be appreciated in the life of the poor, while *iki* applied mostly to the way of life of wealthier citizens.[22]

In general, we can say that in dealing with various ethical problems such as love, the emancipation of the individual, and the writer's mission in society, it was Western ideology and Western literature that became for the majority of Japanese Romanticists a most attractive model. However, this admiration was often noncritical and sometimes drove them to dubious conclusions. In his essay "An Overview of Meiji Literature" ("Meiji bungaku kanken"), Tokoku criticizes the aesthetic ideals of Edo urban literature, contrasting it with the spiritual freedom typical of the culture of the new age, which gives inspiration to the human soul, especially where the ideal of perfect love is concerned. It is this notion of elevated love that Tokoku considered the touchstone of his basic principle: the denial of any compromise in life. Thus, in Tokoku's perspective, the only thing worthy of the highest praise in all of Edo literature would be the double suicide of lovers (*shinju*). According to Tokoku, if there were no *shinju*, there could be no vows of faithfulness in the next world. A passionate and sensitive person, he saw in such a death the climax of love, the triumph of moral strength, and a noble-minded spirituality that could partially redeem the lack of pathos in the image of the lovers.

For Tokoku and his followers, love is a universal creative force that guides man in all his activities. Partially under the influence of such European poets as Wordsworth, the Japanese Romanticists projected "love" as an abstract notion also onto nature, thereby establishing the connection between human and cosmic forces and opposing this harmony to the disharmony of society. Here we find the influence of traditional Zen Buddhist aesthetics, which emphasize the role of a spiritual drive typical of both animate and inanimate objects: "Eternity is in the state of love with the transient world, and this relation between man and the world is Zen; Enlightenment [satori] is the climax of love."[23] In this attitude toward the surrounding world and nature, which combines an ancient tradition with the latest achievements of the Western philosophy of idealism, the peculiarity of Japanese Romanticism is distinctly revealed.

ROMANTICIST QUEST IN THE WORKS OF KITAMURA TOKOKU

Tokoku's overwhelming influence on the poets of Japanese Romanticism can be explained not only by his poetic talent but also by the magnetism of his personality. His short life devoted to the struggle for humanist ideals against philistine morals and his dramatic suicide, which was in itself an ultimate challenge to philistine society, provide the perfect stereotype of the Romanticist poet. His name in Japan evokes the same Romantic image that adheres to the names of Western poets such as Byron, Shelley, Keats, Lermontov, Petöfi, and Poe, all of whom also died young. Perhaps it is this image that enables the influence of such poets to extend not only through geographic space but also through time, allowing them to become intermediaries between cultures and sources of inspiration for poets of later periods.

Scholars have evaluated Tokoku's role in the development of modern Japanese literature in various ways. Some believe that his main merit was his aspiration to embody in literature the concepts of the Movement for Freedom and Civil Rights.[24] Other scholars praise his efforts at deconstructing the traditional Confucian principle of "encouraging good and punishing evil" (*kanzen choaku*) in Japanese literature.[25] Some assert that his major contribution was the promotion of the freedom of spirit and pure love,[26] while still others appreciate his active resistance to the pseudo-classicism of the Kenyusha group.[27] Despite the differences in evaluation, the majority of scholars agree that the "modernization of Japanese verse started with Tokoku."[28]

Tokoku's first serious attempt to create poetry in new forms was a long poem entitled "The Poem of a Captive" ("Soshu no shi"), published in 1889. The poem, which consists of sixteen parts, presents variations on the theme of Byron's "The Prisoner of Chillon." Like Byron's poem, it is structured as a monologue. Critics suggest that the motivation for writing the poem may have been the arrest of some of Tokoku's friends and fellow participants in the Movement for Freedom and Civil Rights. The poem criticizes unjust laws and glorifies the "eternal spirit of the chainless mind," thus echoing Byron's masterpiece. In the poem, the evil the hero confronts takes on a real appearance in the flesh. This evil, however, which is associated with oppression, can be opposed and defeated.

Tokoku's most noteworthy poetic work is his verse drama *The Song of a Magic Land* (*Horaikyoku*), published in 1891. *The Song of a Magic Land* is the exact antithesis to the positive and hopeful "The Poem of a Captive." It is perhaps the gloomiest example of Romanticist dramatic art ever written. Some scholars assert that it has to be analyzed as a free interpretation (*yakuan*) of Byron's poetic drama *Manfred*. There can be no doubt that *Manfred* provided a source for Tokoku's drama. The overwhelming grief and misanthropic moods of the main character, Yanagida Motoo, seem in particular to be indebted to Byron's work. However, it would be incorrect to trace the tendencies found in Tokoku's work to the influence of *Manfred* alone, as the plot and composition have very little in common with Byron's drama.

In *The Song of a Magic Land*, the evil forces the character confronts are of a universal nature, extending to the macrocosm (the devil) and the microcosm (the inner world of the character himself). Yanagida Motoo suffers from the perversity of the world around him, from the dreariness and emptiness of life, and from his own deplorable weaknesses. Motoo is not a Romantic "superhero." His character has a softness and sentimentality, which emerge from the "sensitive emotional outbursts" of Tokoku's personal poetic style.

Nowhere else is exposed with such force Tokoku's inner world, his aspiration for the ideal of "spiritual freedom," his love for mankind and hatred of bourgeois reality, and the incessant torments that led the poet to his tragic end. In his recollection of Tokoku, Shimazaki Toson wrote: "'Rebellious spirit' (*doyo seshin*) is the term that can be applied to Kitamura-kun [Tokoku]. He would drink sake with his friends and then go to write about Shakespeare's dramas, spend all night over articles on modern literature, and then next day go to perform missionary work. That was Tokoku.... In his soul the struggle between deep religiousness and the 'freedom of spirit' never ceased."[29] The echoes of this struggle are heard in *The Song of a Magic Land*.

A passionate Romanticist, a devout Christian, and the author of grief-filled poetry, Tokoku eventually returned to the sources of the indigenous tradition, to the lap of Buddhist beliefs. His theory of the "voices of things" reflects primarily the Buddhist concept that all living things possess a soul and the ability to be transformed after death. Thus, according to Tokoku, animated images of insects and plants can express the sad destiny of humankind best of all. Why insects and plants? Certainly it is because they are the symbols of ephemerality, of the extreme caducity and fragility of human life. The bulk of Tokoku's poems, in fact, represents a version of traditional Sino-Japanese poetry in the style of "flowers, birds, wind, and moon" (*kachofugetsu*), but the poems do have some differences typical only of sentimental Romantic poetry, and they also contain an innovative lyrical note. In fact, they can be regarded as works showing "the soft side" of the gallant paladin of Japanese Romanticism. The Romantic pessimism of Tokoku possessed such magnetism for the literary-minded youth of the time that it became the basis of the Japanese Romanticist school, then emerging around experiments in *shintaishi* verse. Having opened a window into the intimate spiritual world of their leader and guru, Tokoku's late poems, full of the anticipation of death and followed soon by the suicide of the author himself, made a strong impression upon his fellow Romanticists. Now they could appreciate the depth of his words: "In the world of joy there are few people of passion. Passion is a friend of misfortune. Passion is a neighbor of grief."[30]

Tokoku's revolutionary articles on the role of the artist in the new Japan, his heroic drama, and his sweet songs of sorrow and disillusion continued to excite the minds of his followers and admirers for many years after his death, as did his legendary image. His very name became the symbol of Romantic revolt against the new Japanese establishment, against a social order based on the principles of blind loyalty, uncritical faith, and militant nationalist aspirations.

After the death of Tokoku, mainstream Romanticist poetry in Japan, with its slogans of spiritual freedom, universal love, and cosmopolitan aesthetic relativism, became the stronghold of literary opposition to the rising militarist authoritarian regime. As happened several decades earlier in Europe, eventually the Romanticist revolt against the establishment failed, but its ideals survived in the abundant multifaced culture of the

Meiji-Taisho period. Succeeding the Romanticists, all the best literary schools of the early twentieth century took as their starting point the concept of East/West synthesis. Democratic poets, critics, and novelists raised their voices against the ultranationalist ideology promoted by the government. It was the revolt of Art, based on the principles humanism and beauty, against the dark realm of the nationalist "new order." Although in the prewar and war years, military ultranationalist pathos prevailed for a while over the cosmopolitan views of the literati, in works by the early Meiji Romanticists, poets and writers were seeking a source of strength and will for resistance. And it was the same Romantic aspiration for a bright future that helped them build a new country on the ruins of the evil empire.

CHAPTER 8

Diderot's Energistic Philosophy and the Sublime in Evil

Miran Bozovic

The concept of energy is not something peripheral to modern Western thought but has played an extremely important role in its development. Goethe and Nietzsche, for example, readily spring to mind as two key thinkers of modernity who did not take their cue from harmony or order but from energy. Harmony, like order, may be the result of lack of energy. Or to put it another way, there are many different types of order, and those that are bursting with energy are very different from those that lack it. But in the main, energy is no respecter of boundaries and borders, or of harmony or order for the sake of order. The rebelliousness that characterizes so much modern Western thought comes from this basic insight, an insight that may well have been based on the observation that the European social and political order was often rotten and not worth preserving. In keeping with this, it may not be surprising that behind the faith of revolutionaries one can detect the idea that revolutions unleash energies heretofore latent or unknown, energies that will help break apart the encrustations of dead routines and stultifying moral, social, and political orders. Among those writers who stress the importance of energy as a creative force, Denis Diderot, in *Rameau's Nephew*, addresses the issue in the most alarming fashion—by providing the case for the existence of the sublime in evil. It is interesting that the Enlightenment is usually taken as the paradigm par excellence of a model that emphasizes and values rational order, and

Diderot is certainly no less a canonical figure among Les *Lumières* than Voltaire. Usually we associate stress upon darkness and the demonic with the Romantics, but the case of Diderot suggests that revolutionary excess may well be but the excess of the Enlightenment, part of the irruptive energy that was necessary to introduce a new order, a kind of energy that inevitably would manifest itself as crime.

One of the more disturbing and even repulsive traits of the eponymous hero of Diderot's novel *Rameau's Nephew* is the passion and enthusiasm with which he narrates a particularly wanton crime and the undisguised admiration Rameau shows for the perpetrator. While listening to Rameau, his interlocutor—Moi, which is to say Diderot himself—realizes that he finds the euphoric tone with which Rameau narrates the crime as disgusting as the villainy of the criminal himself. When he has heard the story through to the end, he vacillates between the extremes of laughter and rage. He does not know whether to remain in the company of a man he finds increasingly unbearable or simply to get up and leave, and for a moment Diderot feels sick to his stomach. What makes Diderot sick is listening to Rameau discuss a brutal crime the way "a connoisseur of painting or poetry examines the beauties of a work of art, or a moralist or historian picks out and illuminates the circumstances of a heroic deed."[1]

Rameau's story concerns the so-called renegade of Avignon and his rich Jewish friend. The hero of the story is a Christian who pretends to his Jewish friend that he has adopted Judaism, while the latter pretends to have adopted Christianity. On the outside, both men are pretending to be what they are not. But while fear of the Holy Inquisition has compelled the Jew to pretend to adopt Christianity, the Christian pretends to have adopted Judaism only to gain the Jew's trust so that he can rob him of his money. One day, having gained the Jew's complete trust, the renegade appears at his friend's door with some terrible news: an unknown traitor has denounced them both to the Inquisition—him as a renegade who has secretly adopted Judaism while pretending to remain a Christian and his friend as a Jew who pretends to have adopted Christianity while in truth remaining faithful to his religion. Fearing for their lives, the two men decide to charter a ship to flee from persecution. However, on the night prior to their departure, the renegade secretly relieves his sleeping friend of "his wallet, his purse, and his jewels" and sails away with the Jew's entire fortune.

It is at this point that the story appears to reach a climax. This climax, however, proves to be a red herring. Rameau's skillful and cunning narration tricks us into believing that we have reached the end of the story, for we mistakenly assume that we can guess the truth: the renegade simply made up the entire story of the traitor's denunciation in order to fool his friend and make off with his fortune. If this really were the case, then the renegade would have successfully pulled off a clever deception. But his crime would not qualify as a truly "great crime" according to Rameau's high standards. Indeed, if this were all there were to the story, Rameau would most likely think it not worth recounting. However, the story has an appealing and unexpected twist: the denunciation turns out to have been real. Moreover, it turns out that the renegade himself denounced his Jewish friend to the Inquisition. Thus, the next morning, the Jew is arrested and ends up being burned at the stake while the renegade is free to enjoy his friend's riches.

It is in the renegade's gratuitous and entirely superfluous gesture that Rameau sees *le sublime de sa méchanceté*, the sublimity of his wickedness.[2] Although his friend's fortune is already safely in his grasp, the renegade nevertheless denounces the Jew to the Inquisition. Why does Rameau so exalt the sublime in evil? He explains: "If it is important to be sublime in anything, it is especially so in evil"; while people usually "spit on a petty thief," they "can't withhold a sort of respect from a great criminal. His courage bowls you over. His brutality makes you shudder."[3] Further on, Rameau adds that it is the "enormity of the deed" that carries one "beyond mere contempt."[4] According to Rameau, the only way for criminals to avoid being held in contempt is to be consistent in their wickedness, not to vacillate in their principles, and to constantly act in accordance with their nature or character. Thus their conduct will manifest "consistency of character," which is highly valued "in everything"[5]—and therefore also in evil. This is precisely what Rameau admires about the renegade. The renegade could quite easily have made up the denunciation to the Inquisition; this would have allowed him to access the Jew's fortune in a relatively straightforward and elegant manner and, above all, without causing unnecessary harm. However, had he acted in this way, the renegade would not have manifested "consistency of character" and would have remained a petty, insignificant crook whom "nobody would want to resemble."[6] A person who calculates how to carry out a scam in the simplest possible way to avoid paining the victim, without unnecessary harm and so forth, is not genuinely wicked, for he still seems to vacillate in his principles. However, a truly wicked person, a person who is consistent in his wickedness, thinks quite differently. For example, he will think: Not only can I relieve my victim of his money, but I can also hurt him with but a little additional effort on my part and without too great a risk.

Therefore I am not going to pass up a good opportunity to do this as well! Rameau believes it is precisely this "horrible act" that carries one "beyond contempt" and transforms the perpetrator in our eyes from "petty thief" into a "great criminal," from whom we "can't withhold a sort of respect."

The delight with which Rameau relates the renegade's horrible act brings to mind "the aesthetics of murder" developed by members of the Society of Connoisseurs in Murder in Thomas De Quincey's satirical essay "On Murder Considered as One of the Fine Arts" (1827). Just as Rameau discusses a vicious crime "like a connoisseur of painting or poetry examining the beauties of a work of art," so De Quincey's connoisseurs of various methods of bloodshed meet to examine every new murder that has been committed just "as they would a picture, statue, or other work of art."[7] In particular, they examine individual murders with an eye for the good taste shown by the murderer in his choice of victim, place, and time of the murder and with an appreciation of "originality of design," "boldness and breadth of style," and so forth. Based on the above criteria, a De Quincian expert may grade and classify murders on a scale ranging from mere "plagiarism," through crude first attempts that betray the hand of an unpracticed yet promising artist, to the "masterpieces of excellence" and "immortal works" in which individual murderers have carried their art "to a point of colossal sublimity."[8]

De Quincian "artists" are so strict in their adherence to aesthetic standards that if they come to realize that their choice of victim is not going to enable them to achieve the "genuine effects of the art" or that they are going to offend the good taste of the public, they will abstain from the planned murder. De Quincey's example of a murder that did *not* occur as a result of the artist's aesthetic scruples is the planned murder of Kant. At the sight of "the old transcendentalist," the artist changed his mind at the last moment and chose another, much younger victim. Clearly, Kant's prospective murderer was a subtle and sensitive artist who appreciated "how little would be gained to the cause of good taste by murdering an old, arid, and adust metaphysician," writes De Quincey. Since Kant "could not possibly look more like a mummy when dead, than he had done alive,"[9] the would-be murderer most likely concluded that, in this case, his creativity would be altogether wasted.

Let us take a quick look at just one of De Quincey's "principles of murder"[10]—namely the choice of place and time. Although common sense suggests that murderers ought to practice their art after dark and in lonely places, the De Quincian expert would find a murder committed on a moonless night and in a dark lane deeply offensive. Only those murders that depart from this rule count as works of art. Thus, in his "great gallery of murders," undisputed masterpieces include the assassination of Gustavus Adolphus, king of Sweden, who was murdered "at noon-day" and on the battlefield, in plain sight of everyone; and the assassination of a bank porter who was murdered in Edinburgh, Scotland, "in broad daylight" and in "one of the most public streets in Europe."[11] But perhaps the greatest masterpiece of all was the assassination of Jonathan, a high priest who was murdered in the temple in Jerusalem in the middle of Mass, in front of the whole congregation, while the eyes of the entire crowd were upon him.[12] What makes these murders "immortal" is the "bold ... idea ... of a noonday murder in the heart of a great city."[13]

Although Diderot finds Rameau's enthusiasm for the renegade's cruel crime repellent, and although for a moment Diderot even feels sick at the thought of it, he is himself no less fascinated by great crimes and their perpetrators than Rameau. Evidence of this can be found in some of Diderot's ideas on human nature, which he developed in the circle of his closest philosophical friends and which he reconstructs in great detail in his letters to Sophie Volland. Moreover, if Diderot is unwilling to share Rameau's enthusiasm for the renegade of Avignon, this is simply because, in Diderot's eyes, the renegade is not a sufficiently great criminal to deserve his admiration. In one of his letters he sums up a "conversation about human nature" that had taken place the previous day at the dinner table at La Chevrette: "I could not help admiring human nature, even in its moments of atrocity."[14] The conversation developed immediately following a dinner that Diderot had complained had been too heavy for their feeble stomachs. Nevertheless, this did not seem to prevent Diderot from speaking passionately about those criminals who cannot be discouraged from committing their crimes even by the public execution of another criminal, and it did not seem to prevent him from openly admiring great crimes, such as Tarquin's rape of Lucretia. One might conclude that Diderot's stomach was not as weak as he would have us believe.

An example of "human nature ... in its moments of atrocity" that Diderot greatly admired in the conversation is that of thieves who steal from a crowd that is witnessing the public execution of another thief: "Even at the execution of a thief there are thieves at work in the crowd, risking the very same fate they see before them. What contempt for death and for life!"[15]

The act of theft from the crowd, although a considerably lesser crime than murder, could nevertheless still meet the criteria that De Quincey prescribes for the "fine art" of murder. Diderot's thieves show no small measure of good taste in their choice of time and place for carrying out their crimes: they steal "in broad daylight" and in "a public place" before a crowd. This much alone—even disregarding the fact that they are stealing from a crowd that has gathered to witness the execution of another thief—would be enough to lend their thefts a "scenic effect" and to make them quite comparable to the supreme masterpieces in De Quincey's "gallery of murder" (the murder of Jonathan, the high priest; the murder of Gustavus Adolphus, king of Sweden; and the murder of the Scottish bank porter), whose "aesthetic" value derives from their being committed during the daytime and in public places.

Their decision to choose the site of another thief's execution as the scene of their crimes would be extolled by De Quincey as "the beauty of the case" and would be praised by Rameau as an example of "the sublimity of [their] wickedness." This fact alone is sufficient to show that the thieves are not driven to theft by want. Rather, theirs are crimes committed for the sake of committing crimes. Like Saint Augustine, himself a one-time thief in his youth, they take delight not in the things they steal but in the offense itself, in the very act of stealing. When recalling what it was that he loved about his youthful thieving, Saint Augustine says that he loved it "only for the sake of the theft." More than the objects he stole, he loved the theft itself.[16] The same, no doubt, holds true for Diderot's thieves; why else, if not for the love of theft itself, would a thief commit his crime during a public execution of another man for theft? Is it possible to conceive of a more epicurean theft? Could any theft be carried to a point of such "colossal sublimity"? By committing such a pure, motiveless theft, Diderot's thieves would not offend the "severe good taste" of the De Quincian "enlightened connoisseur"; in fact they would most probably compel him to admit that even theft can be a "fine art."

What exactly does Diderot admire about great crimes? As a determinist, Diderot believes "there are and there can be no free beings" and that therefore "there are no actions which deserve praise or blame," for "there is neither vice nor virtue, nothing that must be rewarded or punished."[17] As a determinist, Diderot could hardly be expected to admire the same things about crime that De Quincey does. Even if he did happen to admire the same thing as De Quincey—for example, the boldness of a crime—it would almost certainly be for entirely different reasons. For Diderot, unlike for De Quincey, no action whatsoever—whether a murder or the act of saving someone's life—can constitute a "meritorious performance."[18]

That Diderot is fascinated by something entirely different about the crimes is perhaps best illustrated with a different example of "human nature ... in its moments of atrocity": the example of Robert-François Damiens, who attempted to assassinate Louis XV in 1757, for which he was sentenced to death by quartering. Diderot shows an even greater admiration for Damiens than he does for the thieves at work during the execution of another thief, in spite of the fact that the crime itself *failed*. Diderot first admires Damiens for the "magnificence" of the crime attempted, for he "dared to lift his hand against his king," and above all for the composure he showed on hearing his sentence pronounced. Upon learning of the terrible punishment that awaited him—his flesh would be torn from his body with red-hot pincers, molten lead and boiling oil would be poured into his wounds, and, finally, he would be dismembered by four horses—Damiens supposedly uttered with complete calm, "La journée sera rude"[19] (The day will be hard).

What both of Diderot's examples—the thieves and Damiens—have in common is the occasion of a public execution of a criminal as a motif, as well as the apparent indifference displayed to this by the criminals themselves. In both examples, the criminals appear wholly unconcerned by the execution being carried out. The example of Damiens, presented from the point of view of the convicted criminal, is an amplification of the example of the thieves, for while the thieves merely risk the same fate as that of the thief on the scaffold, a far more brutal version of such punishment inevitably awaits Damiens. Although, strictly speaking, in a deterministic universe there is "nothing which calls for ... punishment," Diderot nevertheless advocates the public execution of criminals. Although anyone can exculpate himself from any act, no matter how horrible, simply by quoting Jacques's words "Can I be anything other than myself, and being me, can I act otherwise than I do?"[20] Diderot also believes that the perpetrators of grave crimes must nevertheless be done away with. *Le malfaisant*, the malefactor, must be "destroyed in a public place" for the "beneficial effects of example."[21] By showing no remorse for his crime in the face of his terrible impending death, Damiens seems to confirm the classical utilitarian rationale that the reformation of criminals is entirely ineffective and that punishment can serve only to set an example to others. For Diderot's thieves, punishment fails even as a deterrent, since they carry on stealing in the face of the very punishment that is intended to deter them from doing so. It may at first appear that in admiring thieves who cannot be discouraged from theft even by a public execution of another thief, Diderot contradicts his own belief in the deterring effects of exemplary punishment. However, this is not the case. In a deterministic universe, one cannot reasonably expect thieves to be intimidated by the execution of one of their peers and to reflect upon their own actions and simply decide to mend their ways. Exemplary punishment might well work in this way for Bentham but not for Diderot. For Diderot, the "beneficial effects" of the deterrent work in a more indirect and roundabout manner. According to Diderot, all people are to a greater or lesser degree *bienfaisant*, beneficent, or *malfaisant*, maleficent; neither are such by choice. The former are simply so because they are "fortunate by birth," while the latter are "unfortunate by birth."[22] *La bienfaisance*, doing good, or *la malfaisance*, evildoing (or doing harm), is not a choice they make; it is simply how they are. Since neither "rational benevolence" nor "rational wickedness" exists,[23] the good deeds of the former do not call for praise or reward, just as the evil deeds of the latter do not call for blame or punishment.

As a rule, characters in Diderot's works all behave in accordance with the above principles. Thus, for example, the hangman in *Jacques the Fatalist*, who saves the eponymous hero's life (it is entirely in the spirit of the novel that Jacques's life is saved by someone who makes his living killing other people), attaches "no importance" to the good deed he performs in saving Jacques's life and expects no gratitude in return. Moreover, the hangman receives Jacques's effusions of gratitude in a "cold and indifferent manner" and even "with contempt," since he is aware that he is "naturally kindly disposed" and has therefore not deserved Jacques's gratitude.[24] Similarly, the blind man in "Letter on the Blind," who has an "extreme abhorrence of theft," does not condemn theft as such but merely opposes it on practical grounds, owing to the "facility with which people could steal from him unobserved, and secondly (still more perhaps) because he would be immediately seen were he to go about filching."[25] In other words, it is not out of respect for other people's property that the blind man finds theft unacceptable but because he himself would make an easy victim of theft and a most incompetent thief. As frightening a spectacle as it may be, the display of exemplary punishment cannot induce those "unfortunate by birth" to reform themselves even if they wished to do so. "Can one cease to be wicked at will?" asks Diderot. He goes on to answer: "Once the crease is made, the fabric remains creased."[26] Since our will is nothing but "the final result of everything which one has been since birth right up to the moment where one is,"[27] and since our every volition is determined by who we are and our personal history at the moment we act, the most that can be achieved by punishing someone is to insert a new link in the "chain of causes and effects" that makes up his life and that may influence his behavior and determine his will in the future. In other words, punishment at best "corrects" or "modifies" the wicked; it cannot convert or transform them into virtuous persons.

Both of Diderot's examples of "atrocious" human nature are rhetorically effective. In selecting these examples, it would appear that Diderot is primarily motivated by "principles of taste." In his letters, he writes about them with undisguised enthusiasm and with a keen ear for the grisly details that ranks alongside Rameau's appraisal of the originality to be found in degradation and De Quincey's aesthetic treatment of murder. Even De Quincey's "advanced connoisseur" would blush at the tone and words with which Diderot writes about them in his letters. In short, Diderot himself seems to treat a criminal act "like a connoisseur of painting or poetry examining the beauties of a work of art." Both Rameau and De Quincey would have been horrified by Diderot's belief that "great crimes" restore our faith in humanity and that as long as thefts are still committed at the foot of the very scaffolds on which thieves are punished, and as long as people like Damiens exist, then there is still hope for humanity.

Before examining why Diderot admires "great crimes" and their perpetrators, it would perhaps be useful to briefly consider what sort of people arouse his pity or contempt. The people he pities or scorns are those who "cannot choose either vice or virtue," those who "are incapable of either immolating other people or sacrificing themselves," and those who are "unhappy" irrespective of "whether they act well or badly"[28] (in other words, those who, as Rameau would put it, are "equally inept at good or evil"[29]). Such people, says Diderot, are "good for nothing"; they make neither good honest men nor good scoundrels. What is striking is that Diderot scorns not only those who have not chosen virtue, those who are incapable of sacrificing themselves—that is, those who are inept at good—but also those who have not chosen *vice*, those who are incapable of immolating *others*—that is, those who are inept at *evil*. Just as on the one hand he pities and scorns both those who are inept at doing good and who at best make only mediocre benefactors, and those who are inept at evil and therefore make only mediocre criminals, so too, on the other hand, does he admire both the great benefactors and the *great criminals*. As a rule, we tend to admire the benefactors and scorn the criminals. The greater the benefactor, the more we admire him; the greater the criminal, the more we scorn him. By contrast, Diderot consistently scorns all those who are petty or mediocre, regardless of whether they are such at good or evil, and he consistently admires all those who are great, regardless of whether they are such at good or evil.

In addition to scorning petty or merely mediocre criminals and admiring great benefactors—attitudes that most of us would readily share—how is Diderot also able to scorn petty or mediocre benefactors and admire great criminals? When speaking of his hatred for "petty base actions" (for example, the vicious gossip mongering of otherwise pious women or the occasional tyranny of otherwise benevolent masters), he justifies his attitude toward "great crimes" by saying: "I hate all those petty base actions that reveal nothing but an abject soul, but I do not hate great crimes: first, because they are the stuff of beautiful paintings and beautiful tragedies; and also, because great and sublime actions and great crimes embody the same character of energy. If one man were not capable of setting fire to a town, another man would not be capable of leaping into a chasm to save him. If we could not have had a Caesar, neither could we have had a Cato."[30]

Similarly, in his letter to Sophie Volland, Diderot rationalizes his admiration for thieves: "If the wicked did not have such energy for crime, the good would not have the same energy for virtue. If a debilitated man has lost the strength to commit great crimes, he will have no strength for great acts of virtue.... If Tarquin no longer dares to violate Lucretia, Scaevola will not hold his wrist over the burning coals."[31] And he justifies his enthusiasm for Damiens by saying, "There are no powerful actions in weak nations. A sybarite is as incapable of murdering his neighbour as he is of saving his mistress from a fire."[32]

Given his utilitarian theory of punishment, that perpetrators ought to be publicly punished to set an example, it is clear that Diderot does not admire great crimes as things in themselves and that, unlike Rameau, he does not aestheticize evil. Instead, Diderot's rationalizations share a common amoral, or rather premoral, concept of "energy,"[33] which when liberated brings about both *les grands biens*, great goods or great acts of virtue, and *les grands maux*, great evils or great acts of wickedness. When this energy is restrained, however, the best we can hope for is a mere mediocrity of both good and evil. Leibniz's influence here is clear, not only in the vocabulary (*les grands maux* and *biens*) but also in the specific examples of "great evils" (for instance, the rape of Lucretia), although Diderot's "economy" of evil is rather different and perhaps less "admirable" than Leibniz's. For Leibniz, God permits evil because at some point in the future it will inevitably give rise to an incomparably greater good—a good that in the absence of this evil would not have come about.

Thus, for instance, God permitted Tarquin's rape of Lucretia because he knew that this crime would serve "for great things": it was precisely this crime that led to the founding of "a great empire," which provided mankind with "noble examples."[34] If this crime had not taken place, the greater good itself—that is, the "great empire," "noble examples," and so on—would also have not occurred. The "great empire" and "noble examples" constitute that "greater good" through which Tarquin's crime is "recompensed with interest."[35] Although Diderot believes that if Tarquin had not committed the crime, Scaevola would not have plunged his right hand into the flames, he does not regard Scaevola's noble act of courage as constituting a "greater good" that in any way serves as recompense for the rape of Lucretia (in spite of the fact that, historically, Scaevola's gesture is perhaps even more closely and immediately linked to the rape of Lucretia than the "great empire" and the "noble examples"). In Diderot's view, the energy manifested in "great crimes" or "great acts of wickedness" is merely a guarantee that a no less magnificent energy is also possible "in virtue" and that sooner or later this energy will manifest itself in "great and sublime actions" or "great acts of virtue." For example, Tarquin's boldness in evil guarantees the possibility of an equal enthusiasm to perform a good act. Likewise, the existence of Caesar the dictator guarantees the possible existence of the noble Cato of Utica; the zeal to commit a crime guarantees a comparable zeal for virtuous acts; the existence of irredeemable evildoers such as Damiens, who do not fear death, guarantees the possible existence of heroes such as Regulus, who are equally fearless in the face of death for a good cause; the existence of people who are capable of ruthlessly murdering their neighbors guarantees the possibility of people brave enough to save their fellow man; and so forth. Since all of these acts are governed by one and the same energy, which manifests itself sometimes in evil and sometimes in good, it is impossible to have one without the other. If one wishes to have people such as Scaevola, Cato, and Regulus, one must also accept people like Tarquin, Caesar, and Damiens. If one is not willing to accept "great crimes" but instead prefers "petty base actions," then one cannot hope for any "great and sublime acts" of virtue either. Diderot celebrates "great crimes" because their occasion reasonably anticipates "great acts of virtue." Likewise, he celebrates the corruption of morals because moral decay guarantees that moral growth and flourishing are also possible. When one of the participants in the conversation laments the rise of infidelity, deception, and hypocrisy in today's society, Diderot says, "If there is really more cheating, more double-dealing and more debauchery nowadays than ever before, then there is also more sincerity, more integrity, more true affection, more feelings, more sensibility and more enduring passions than in previous times."[36] That is to say, the existence of adulterers and debauchees guarantees the possible existence of truly devoted and loving persons. Such an "economy" is at work in "all other things." No matter which activities people engage in, the existence of people "who will perform those activities poorly" is always a guarantee of the existence of people "who will perform those activities well."[37]

Diderot does not admire thieves because they are original in their degradation, as Rameau would say, or for "strength of conception" or "originality of design," which De Quincey would celebrate, but rather for the energy they display in their conduct. This energy is so great that even the threat of the death penalty cannot restrain it. Moreover, for Diderot, this energy is so magnificent that it compels thieves to risk the death penalty at the very time and place of a public execution for theft, which is supposed to serve as an example and intended to discourage them from the very acts they are committing. Diderot does not admire Damiens for his witty remark—although his remark is a perfect example of gallows humor and as good as any recounted by Freud in his *Jokes and Their Relation to the Unconscious*[38] —but rather for the energy that in this case is even greater and that cannot be constrained by the prospect of an inevitable death or the fact that he himself will soon embody one of those horrific deterrents intended to serve as an example to others. It is precisely this colossal energy—energy that cannot be constrained by even the prospect of total destruction—that makes Damiens and the thieves "great criminals" in Diderot's eyes, even though the latter are merely crooks who rifle through the pockets of the curious crowd, while the former is a failed assassin who botched a regicide. For Diderot, the incomparable energy of Damiens and the thieves bespeaks a certain "greatness of soul." A less impressive energy, however, is manifested in the sly deception of the renegade who hypocritically robs his friend and then makes sure that he falls into the hands of the Inquisition while he himself makes a cowardly escape. For Diderot, such an act can bespeak only "an abject soul."[39] Indeed, the contrast between the renegade of Avignon and Diderot's thieves could hardly be greater: to rob a sleeping friend in the night within four walls is one thing—any villain, no matter how "inept at ... evil," could accomplish it; a theft carried out in broad daylight among a crowd witnessing the execution of another thief is quite another. Is it any wonder then that Rameau's admiration for the renegade makes Diderot sick to his stomach? Despite the atrocious nature of the crime, and despite the fact that it is successfully and impeccably performed, the renegade's act is a "base action" rather than a "great crime." The renegade is great only in the eyes of someone who, like Rameau, has himself failed in evil. Although Rameau elevated depravity and wickedness to an "art form" and an ethical attitude (a typical maxim of Rameau's ethic would be: Always repay the good someone has done you with an evil [40]), he never accomplished a truly great evil. Instead he remained a beggar to the last, forced to pimp his own wife. It is precisely because Rameau is so "inept at ... evil" that he speaks "with the most profound admiration" of all those who are "more perfect" in evil than himself.[41]

At first glance, Diderot's two examples of "atrocious" human nature may seem ill suited to illustrate his concept of energy, since they arouse admiration entirely for the wrong reasons. However, a closer examination shows that Diderot has chosen his two examples wisely and with the utmost care. Diderot must have known that he would be hardly likely to win over the advocates of free will to his premoral concept of energy by illustrating his argument with "great acts of wickedness," such as Tarquin's rape of Lucretia. On the other hand, however, he knew that it would be quite possible for the advocates of free will to admire the energy displayed by the thieves and by Damiens—in spite of the fact that this energy manifests itself "in crime"—although, as advocates of free will, they would be unwilling to share the determinist's enthusiasm for most other "great acts of wickedness" despite the imposing energy manifested in them.

CHAPTER 9

Saintly Rebels: Gandhi, the Emir Abdel Kader, and the Philosophy of Positive Passivity

Waddick Doyle

In considering the opposition between revolt and spontaneity on the one hand and harmony and order on the other, through East/West perspectives, I argue in this paper that two non-Western traditions, Hinduism and Islam, provide means to think differently about these oppositions. French scholar René Guénon argued that East and West were as much symbolic as geographical categories,[1] with East representing cultures where the metaphysical dominated the material. It would be hard now to see any part of today's world that is not Western in Guénon's terms.

In this paper, I wish to consider two emblematic figures of revolt from the East, the leader of the Indian anticolonial struggle Mahatma Gandhi and the emir Abdel Kader, leader of the Algerian resistance against French colonialists in Algeria. They both provide grounds for separating order and harmony as two distinct and often opposing qualities. Both provide grounds to consider that revolt and spontaneity are means of restoring harmony. This was particularly the case when societies based on religious traditions aspiring to peace were invaded by Western powers. However, both Mahatma Gandhi and the emir Abdel Kader were also key figures in the dialogue between civilizations. Both studied Christianity and saw its common points with their own religions, and both argued for a commonality of religious practice: Gandhi was famous for his openness to Islam and Christianity, while the emir Abdel Kader argued that Christianity and Islam were essentially compatible and that their differences could potentially be resolved, but there was too little will to do so.

In both cases, their anticolonial resistance was inspired by traditional Eastern ways of thinking. How did anticolonial struggles based on non-Western religious and spiritual concepts conceive of their revolt? Both Gandhi and Kader saw their revolt as restoring a harmony of the traditional order that had been destroyed by colonial aggression. This harmony was essentially to do with techniques of the self and ethics. Gandhi and Kader were consummate political leaders of revolts, but both were and still are also considered saints by their followers, and in some cases by some of their colonial enemies. They combined action and contemplation, harmony and revolt, self-cultivation and politics, activity and passivity creating a harmony of "opposites."

Modernity's intimate connection with the European colonial project has still not been sufficiently explored. It is characterised by the Western liberal order (or disorder) being imposed upon the East and the South of the world despite unsuccessful resistance. Syed Hossein Nasr has argued that philosophically this became possible because Descartes and others proposed a radical discontinuity between human consciousness and nature. Science then developed modern technologies dedicated to the domination of nature.[2] The conquest and domination of the non-West was linked to this larger project of domination, partly in pursuit of natural resources and partly because the world and other cultures could be conceived of as nature to be dominated and developed.

The subjugation of traditional cultures all over the world by modern imperial industrial nations—whether the conquest of Algeria, the occupation of Congo, or the Chinese Opium Wars—was achieved by violence. The anticolonial movements of the twentieth century found inspiration in Western modernist ideas such as Marxism and nationalism. Gandhi and long before him Abdel Kader were exceptions because they formed a resistance born from traditional spiritual doctrines about the nature of reality and human subjectivity. We will argue that the doctrines of Sufism and Yoga have similar conceptions of freedom and indeed of the human subject. Both argue for a continuity between knowing and knower, defining freedom as the experience of truth conceived of as nonduality or unicity. That is, both reject the notion that the human is distinct and autonomous from the nature of being that all life shares. Hence revolt in both cases is the attempt to restore the necessary conditions for freedom and the harmony of unicity.

GANDHI

Gandhi, the emblematic figure of the Eastern anticolonial struggle, termed his political philosophy Satyagraha, which is normally understood as nonviolence. In his own words

> I have also called it [the] love-force or soul-force. In the application of Satyagraha, I discovered in the earliest stages that pursuit of truth did not admit of violence being inflicted on one's opponent but that he must be weaned from error by patience and compassion. For what appears to be truth to the one may appear to be error to the other. And patience means self-suffering. So the doctrine came to mean vindication of truth, not by infliction of suffering on the opponent, but on oneself.[3]

Satyagraha literally means "holding to the truth" (*sat* meaning both being and truth). Thus truth in the sense of *satya* is defined as being. It is not a theory of correspondence between an external empirical reality and its representation in language. Truth is associated with soul-force by Gandhi because the word *atman*, the soul or self, is etymologically derived from the term *breath*, which refers to the self in continuity with universal being (*brahman*), just as breath is the same as air. Truth and soul-force are associated with patience and endurance because they allow the human subject to be passive and receptive to being. This notion of truth is in turn associated with love because love is associated with the human self or soul losing its sense of individuality in the overall unicity of being represented by *brahman*. Hence the whole concept of Satyagraha is one about the self's continuity with being and the assertion of that being in the face of injustice.

While nonviolent resistance was generally understood in the West as the refusal to take up arms and accepting the blows of the other without responding, according to Gandhi it was in fact a technique of being present in any particular context, of asserting the subject's being against those who would deny it. Holding to being and witnessing are the key notions. The human subject's knowing consciousness is seen as an emanation of the universal consciousness of being. In classic Hindu doctrine,[4] human consciousness of the self is linked to a consciousness of a greater self. Moving from one form of consciousness to another is a key element in the subject's development and taking on power. Faced with aggression, the subject holds to truth and asserts being. While this process may appear passive, it is in fact an active process.

Gandhi would practice Satyagraha in the famous 1930 Salt March confrontation, where the British claimed to have a monopoly on the right to make salt. Gandhi and his followers refused to leave the salt works, taking blows from the British but not returning them. But nevertheless they continued to assert their right to make salt. Strategically, the British could have chosen to ignore, kill, or arrest Gandhi, but in any of these cases; he would have become a symbol of resistance to their power while exposing the British willingness to use violence to maintain the salt monopoly. Gandhi ensured that his passivity in the face of violence became a symbol of active resistance. His passivity to being was such that he was able to be an incarnation of harmony through the practice of patience as well as an active resistant to British power.

Nonviolence, or *ahimsa*, is often misunderstood. It is the *summum* of Yogic morality, the principal *yama* or moral obligation of the eight *yamas* and *niyamas* (something resembling the Ten Commandments).[5] As well as being a moral position and attitude, it is also a psychological position that seeks not to have malevolence toward others. It is also a philosophical position about restoring and experiencing unity and harmony between self and others as well as self (*atman*) and the universal self (*brahman*). Indeed ahimsa, or not causing harm, involves the conception of not causing or wishing harm with the mind. This attitude seeks to develop the force of presence and is certainly never a willingness to give in to cede power against a more powerful violence. It is an activity of consciousness that appears to be passive but must not be confused with cowardice. Gandhi made this abundantly clear on many occasions:

> I do believe that, where there is only a choice between cowardice and violence, I would advise violence.... I would rather have India resort to arms in order to defend her honour than that she should, in a cowardly manner, become or remain a helpless witness to her own dishonour.... But I believe that nonviolence is infinitely superior to violence, forgiveness is more manly than punishment.[6]

> My creed of nonviolence is an extremely active force. It has no room for cowardice or even weakness. There is hope for a violent man to be some day non-violent, but there is none for a coward. I have, therefore, said more than once ... that, if we do not know how to defend ourselves, our women and our places of worship by the force of suffering, i.e., nonviolence, we must, if we are men, be at least able to defend all these by fighting.[7]

Gandhi does not reject violence as the worst evil. He considers it superior to cowardly submission. Indeed nonviolence is possible only when a possible violence is renounced. In this simple hierarchy of effort and opposition, there are three forms. The lowest form can be seen as cowardice, the middle form as violent and courageous violent resistance, and the highest form as nonviolent resistance or the assertion of being or harmony. It may be helpful to explore Indian philosophical conceptions of harmony, violence, and inertia and the relations between them and how these ideas motivated and formed the basis of Gandhi's anticolonial struggle. These distinctions, however, are based on the ancient categorization of all transient phenomena in Indian philosophy.

Samkhya and Yoga philosophy and also chapter fourteen of the *Bhagavad Gita* explain clearly the doctrine of the three *gunas*.[8] All material or more precisely transient phenomena are a combination of three modes, or *gunas*, in a variety of different manners. These terms can be used to describe human behaviors, food, emotions, objects, energies, and so on. They consist of *tamas*, which is inert, dark, and based on the past; *rajas*, which is based on action, effort, violence, passion, and speed; and *sattwa*, which is based on harmony, clarity, and being. The relationship between them is hierarchical, with *tamas* being the inferior. Even the highest of these, *sattwa*, is not considered truly spiritual as it is part of the transient or material world of *prakriti*, which contrasts with the absolute called *purusa* in the Yoga school or *brahman* in the Vedantic school. Put simply, revolt is *rajasic* action, and *sattwa* corresponds to harmony. Gandhi's classification hence rejects *tamasic* inert cowardice in favor of *rajasic* self-defense but privileges the harmonious *sattwa* as above *rajasic* revolt.

This philiosophy makes a distinction between two types of passivity. It is important not to confuse the harmonious passivity *of sattwa* with the inert lazy passivity of *tamas*. If *rajas* is passion or desire, *tamas* is repressed desire or the incapacity to form passion or desire, a state of apathy, and *sattwa* is the capacity to be serene to sublimate desire and passion into a harmonious state. While Indian philosophy suggests the cultivation of *sattwa*, it argues that the *tamasic* and the *rajasic* are always necessary. Serene passivity or *sattwa* is where the subject is calm enough for *purusa* (absolute being) to shine through, and this is the opposite of inertia. In one state the subject is dominated, and in another it is beyond domination.

If in an individual the *tamasic* element is strongest, he will be dominated by habit, laziness, and cowardice. The will of the individual is not formed and is often subject to the domination of others. Human will is formed through effort or *rajasic* elements suppressing *tamas*. *Rajas* allows the formation of desire and its realization. The *rajasic* individual can be angry and refuse domination but must seek action and effort. Krishna tells Arjuna in the *Bhagavad Gita that* he must make an effort and go to war against his own brothers to restore the dharma: the law, order. War involves the courage of being prepared to die and to kill. However, this *rajasic* nature can be transformed into *sattwa*, says Krishna to Arjuna in the *Bhagavad Gita*, through karma Yoga, acting purely out of duty and seeking no profit from one's actions. As this happens, the human being becomes transparent to truth, *satya*. Gandhi called this practice Satyagraha, translated incorrectly as "nonviolence."

According the *Bhagavad Gita*, the warrior, then, must be prepared for surrender to the absolute (*isvara pranidhana*) through *sattwic* practices of detachment. The sensation of being dominated by another through his capacity to provoke fear is an example of *tamas*, but the desire for revenge and victory and even the will to dominate is *rajas*. These are distinct from *sattwa*, which is the capacity of the subject to assert being and harmony, including in the face of violence, and to take power beyond the level of conflict. Hence pardon and mercy are *sattwic* in their nature. The *sattwic*, however, goes beyond simple forgiveness and becomes an effective way of assuming power, of generating autonomy. The subject is formed through courage and developed through detachment.

When detached from seeking a strategic outcome from his action and attached to the absolute, the anticolonial subject cannot be dominated with difficulty, despite being nonviolent. The s*attwic* becomes not only a means of self-improvement but also a means of asserting freedom in the face of oppression. Negative freedom (being free of external restrictions) and positive freedom (the capacity to achieve) become linked to a notion of freedom as the experience of presence or nonduality. Although it appears to be passive, it is very powerful. Nonviolence is not only an ethical stance;

it is a cognitive technique that disarms its opponents and actually could be defined in Nietzschean terms as a will to power.

Gandhi himself constantly referred to this notion of detachment from outcomes: "By detachment, I mean you must not worry whether the desired result follows from your actions or not so long as your nature is pure, your ideas correct."[9] Hence the technique of Gandhi was to act while being detached from the result of action. Attachment to the result is *rajasic*. Gandhi, however, argued that one should not be indifferent in a *tamasic*, apathetic manner: "But this, in no way, means indifference to the result. In regard to every action, one must know the result that is expected to follow, the means thereto and the capacity for it."[10]

Gandhi suggested that one should behave strategically but not invest one's being and identity in the result. This detachment could be realized only by going beyond harmony and action. Gandhi argued that nonviolence is *sattwic* and that cowardice is *tamasic*. It is easy to confuse *tamas* and *sattwa*, as both appear to be passionless (*tamas* being unfulfilled desire and *sattwa* being renounced desire). *Tamas* is apathetic whereas *sattwa* is serene.

Richard Attenborough's film on Gandhi's life brought out these notions of positive passivity very clearly. In the scene where Gandhi was attacked by the cavalry in South Africa, his technique consisted of lying on the ground in front of the horses, making the horses afraid and unable to go on. This illustrates almost literally the *sattwic* technique of asserting being against action and the superiority of being and patience over violence and the will to dominate.

This sense of asserting being to reestablish power, Satyagraha, "holding to being," was the essence of both the technique and the ethics of nonviolence. It is an assertion of being, the practice of patience, an actively chosen passivity in the face of the other's attempt to dominate. Patience, the capacity of being to endure in the face of *rajas*, is the opposite of *tamas*, which is the incapacity to act, with one becoming an object and accepting one's objectification. *Sattwic* passivity is more powerful than *rajasic* activity, but both involve the subject becoming more fully itself.

Revolt viewed from a Gandhian or Yogic position, then, is an attempt to restore a harmony of subjectivity. In Yogic terms, the human being in a *tamasic* state is an object of his own habits and fears and is dominated by others. The individual dominated by *rajas* seeks to assert power by rejecting the domination of others and a will to power. The *sattwic* individual is able to refuse the domination of others without seeking domination. He seeks illumination of the subject through a capacity to refuse to act, to refuse to obey. In doing so, the subject becomes not only someone who revolts but someone whose resistance becomes the sign of a rejection of a whole axiology. This is clear in the case of Gandhi's trial, where he revealed the hollowness of the colonial ethic and power structure. Gandhi, when accused of sedition, responds to the magistrate,

> Non-violence is the first article of my faith. It is the last article of my faith. But I had to make my choice. I had either to submit to a system which I consider has done an irreparable harm to my country, or incur the risk of the mad fury of my people bursting forth when they understood the truth from my lips. I know that my people have sometimes gone mad. I am deeply sorry for it; and I am, therefore, here to submit not to a light penalty but to the highest penalty. The only course open to you, Mr. Judge, is, as I am just going to say in my statement, either to resign your post or inflict on me the severest penalty.[11]

This is a rejection of not only the whole English legal system in India but also the axiology on which it is based, and hence it is a revolt. But it is also a technique for establishing his moral authority through his *sattwic* indifference to a prison sentence, such that he impressed even those whose role was to dominate him. Strangman's account of the trial reports the judge saying,

> The law is no respecter of persons. Nevertheless, it would be impossible to ignore the fact that you are in a different category from any person I have ever tried or am likely ever to try. It would be impossible to ignore the fact that in the eyes of millions of your countrymen you are a great patriot and a great leader; even all those who differ from you in politics look up to you as a man of high ideals and of noble and even saintly life. I have to deal with you in one character only.[12]

The English sought to make the Indians their subjects through domination of violence. Gandhi rejected that domination through the passivity of his being, reestablishing order and harmony by referring to a morally superior traditional order. Hence Gandhi caught them in a paradox: he became a sign of a superior moral order and of their inability to dominate him even while imprisoning him. In some ways, the Gospels' descriptions of the trial of Jesus by Pontius Pilate reminds us of this—a harmonious indifference that rejects the moral basis of the authority judging him.

The recognition of Gandhi by the judge as saintly ushers in a new moral order as announced rebellion is linked to harmony. Gandhi is a passive person who accepts the court but refuses to accept its authority. However, his very existence, the presence of his being witnessed by others, became symbolic of active resistance.

The cultivation of *sattwa* led to a point where he turned himself into someone who not only witnesses but is also witnessed, and hence he became an object of admiration, a symbol of the failure of the British to dominate him.

Again, the British were caught in a double bind. If they killed or imprisoned him, he would become a sign of resistance to them and a martyr. If they let him act, his defiance toward them would become a symbol of rebellion. Semiotically, he transformed himself into a sign, which the British could not suppress—that is, he became an irreducible sign. He appeared spontaneous to them, unpredictable, for he was not motivated by the fear or desire that his oppressors interpreted as belonging to his own interests. (Spontaneity in an Indian context can be understood in this way; the man who is free of *rajasic* desire is detached and thus can be spontaneous and unpredictable.)

In this process, photography, film, and the press are very important. Gandhi constantly assured that his acts of nonviolent defiance were witnessed by the media system of his day. He wished to be witnessed. Although this may not have been intrinsic to his struggle, witnessing of the endurance of his presence was essential to this process. He moves from a passivity of action to an activity of presence, which is the essence of the restoration of harmony and order.

In Yogic thought, when the appropriate balance of the three *gunas* is achieved, the presence of *purusa* becomes evident, manifest. Put another way, this occurs when the human subject behaves without fear and desire and lives in the present. Hence spontaneity is linked to harmony, as it is seen as harmony with presence; this is not strategic thinking but spontaneous action embodied in the ethical being of the individual. At this point, this individual's presence becomes itself a witness to being. In Yogic philosophy, freedom is the capacity of the subject to live without predications and hence to be transparent to absolute being.[13] The understanding of the three *gunas* enables us to reconsider notions of passivity and to conceive of the relations of harmony to revolt in the anticolonial struggle.

LEVINAS

Of course, Gandhi's notion of violent resistance and his ethical views are not unique. Indeed, we can think that this way of action or thinking is not exclusively Indian, and we can find other examples of this thinking. Passivity to one's own subjectivity can become the capacity to act in the world. Revolt and spontaneity are then seen as intrinsic to genuine harmony and thus not opposed to it. The *rajas* of passion is transformed by patience, and at the cusp point between the two lies the human will and its formation.

In a Western context, this force of patience has been asserted by the Franco-Lithuanian philosopher Emmanuel Levinas in *Totality and Infinity*. Coming from a Jewish tradition, Levinas made a similar point to Gandhi. He argues that the patience and the capacity of the subject to endure, to witness itself in the face of domination, is the source of human freedom. Drawing upon Hegel's master–slave dialectic of *The Phenomenology of Spirit*, Levinas argues that the one who seeks to dominate has a weakness in that he wishes the dominated person to recognize him as dominant. Hence even in the most extreme circumstances, such as those Jews experienced in the Second World War, there remains the last freedom, which is to refuse to recognize domination. This freedom remains in consciousness, says Levinas:

> In suffering the free being ceases to be free, but, while non-free, is yet free. It remains at a distance from this pain by its very consciousness, and consequently can become a heroic will. This situation where the consciousness deprived of all freedom of movement maintains a minimal distance from the present, this ultimate passivity which nonetheless desperately turns into action and into hope, is patience—the passivity of undergoing, and yet mastery itself. In patience disengagement within engagement is effected … in this extreme consciousness, where the will reaches mastery in a new sense, where death no longer touches it, extreme passivity becomes extreme mastery. The

egoism of the will stands on the verge of an existence
that no longer accents itself.[14]

The similarity between Levinas's and Gandhi's conceptions of the irreducible freedom of consciousness would indicate that such a notion of patience goes well beyond the Indian universe. The oppressed does not become the object of his domination so long as patience allows his consciousness to exceed his domination. Just as in the case of Gandhi and his followers lying down in front of the colonial cavalry, mentioned above, becoming passive and patient can turn the energy of the aggressor back toward himself. The dominator is often a colonial subjugator who seeks to force the colonial subject to recognize his own subjugation. The last moment of freedom is this capacity to assert being. It is a refusal to recognize domination. Comparing Levinas and Gandhi has allowed us to consider that there are different ways of conceiving harmony, revolt, and spontaneity. The human subject has the capacity to achieve a type of mastery through patience, which establishes harmony and order.

EMIR ABDEL KADER

Islam proposes a theory of jihad, which has often been interpreted as war and the opposite of nonviolence.[15] Yet if we consider one of the first anticolonial resistance movements, that of the Algerian emir Abdel Kader, who took up arms against the French invasion in the 1830s, we can see some similarities to Gandhi, particularly in the way that revolt is conceived in relation to harmony. Abdel Kader's Islamic philosophical heritage was far from Gandhi's, yet like Ghandi his revolt was drawn from the philosophy of his religious tradition. His election as emir by the Algerian tribes to lead the revolt against the French invasion came as a surprise, as he was a mystic and an intellectual who wrote on Ibn Arabi and not a general or politician. Abdel Kader was an intellectual, a freedom fighter, and also a Sufi, and he believed in the absolute unity of being (*wahdat al-wujûd*) and the transcendent unity of religions (*wahdat al-adyân*).

The doctrine of the absolute unity of being (*wahdat al-wujûd*)[16] implies the same continuity between the subject and an underlying absolute reality as we saw above with Yogic philosophy. Abdel Kader also believed in the capacity of the individual subject to realize this experience of unity by becoming dead to the material world. Indeed, Abdel Kader once told Dupuch, the bishop of Algiers, that all his anticolonial activities were from a sense of duty to defend his country to follow the example of the prophet Muhammad, whereas he felt dead to the material world.[17] What is similar to Gandhi is the idea of privileging a type of passivity, which meant that he was unable to be dominated by others because he neither desired nor feared anything and thus could not be manipulated.

Abdel Kader led the armies of the Algerians against the French after they invaded Algeria. The French behaved with brutality, using rape, genocidal killings, and other atrocities, a policy that was described at the time by Bopichon as "terrorism," in one of the first uses of the word.[18] Shah-Kazemi quotes an 1883 government commission: "We massacred people carrying [French] passes, on a suspicion we slit the throats of entire populations who were later on proven to be innocent; we tried men famous for their holiness in the land, venerated men, because they had enough courage to come and meet our rage in order to intercede on behalf of their unfortunate fellow countrymen; there were men to sentence them and civilized men to have them executed."

In the face of such barbarity in his enemy, Abdel Kader became famous for his magnanimity, generosity, excellent treatment of prisoners, and openness toward Christianity. He fed prisoners when he had barely enough to eat himself, and he punished any of his men who treated prisoners badly. Visiting clergy were surprised to see that he made sure the French were able to practice their religion. The French signed treaties that they did not respect and broke their word, finally sending him as a prisoner to France in 1847, despite having promised him safe passage to Istanbul. However often they broke their word, Abdel Kader remained constant to his. Like Gandhi, Abdel Kader was a figure of such moral and spiritual value that he became a witness to the harmony of a traditional order in the face of the modern colonialist strategic one. In exile in Damascus, he protected Christians from Muslims and Druze trying to massacre them. He was admired greatly by many French as well as Muslims.

The notions of revolt, of war and activity, for Abdel Kader were very important. This were encompassed in the concept of jihad, or effort. Hence for Abdel Kader, war against the invader constituted a required effort, as did chivalrous conduct, or *futuwah*, toward his opponents.[19] These ethical codes involved the individual doing ethical work on the self—following codes of good behavior and restraining any desire for revenge. Gandhi was famous for his saying "An eye for an eye makes the whole world blind,"[20] and Reza-Kazemi's discussion on Abdel Kader in the article quoted above is entitled "Not an Eye for an Eye."[21] Both figures demonstrated the virtues of mercy and pardon in their revolts and demonstrated that revolt can be conducted in a way that creates harmony between opposites in both its execution and its result.

For Abdel Kader, this ethical activity was minor and a prelude to what was called jihad *kebir*, or the great jihad or struggle against the ego. The minor jihad was that of war against the oppressor, but just as in the Indian anticolonial struggle, the real work consisted in detachment from the result of one's actions and acceptance of the will of God. In many ways he embodied the Koranic verse 5:7: "Oh you who believe be upright for God, witnesses in justice and let not hatred of a people cause you to be unjust. Be just that is closer to piety." For his part, Abdel Kader was to become increasingly contemplative. Indeed, this was so much the case that Leon Roche, a French officer sent to spy on him, came back reporting that after fighting all day, he spent the nights in contemplation and that he could be compared only to Christ.[22] This passivity of contemplation, of making the individual a reflection of the luminous fundamental nature of reality, allows us to reconsider the concept of revolt. Revolt may be against an oppressor, but it is fundamentally linked to harmony when it is combined with revolt against first the ego and then the colonial oppressor.

We may consider the relations between revolt and harmony in a different light if we recognize that some precolonial societies privileged the ethical transformation of self by contemplation over strategic action and that the colonial order threatened this with a philosophy of ego satisfaction and ambition. Just as Yogic philosophy uses the three *gunas* to explain the differences between activity and passivity in the transformation of the subject, Islamic philosophy has a typology of different aspects of the self that allows us to understand passivity and activity. This is articulated by Abu Talib al Maki. Like the term *atman* in Sanskrit, the term *nafs* is ambiguous in Arabic, but it refers to the self, sometimes as ego and sometimes as *al-nafs al-ammāra bi l-sū'*—the self that is subdued entirely to desire and passions; as *al-nafs al-lawwāma*—the self that is reflexive about its ethical choices and is conscience; and as *al-nafs al-mutma'inna al-mardiyya al-rādiya*, where the *nafs*, self, is at peace.[23]

Although these categories do not fit perfectly with the three *gunas*, they do provide a hierarchy of activity, where passivity indicates the lowest and highest states and where contemplation is seen as the fruit of action. We find two types of passivity: being passive to emotions and desires, and being passive to an underlying reality. This ethically ordered theorization of subjectivity allows us to see arguments similar to those of Gandhi. Revolt is necessary to allow conditions for the reestablishment of peace and harmony under the leadership of those who have succeeded in transforming their own natures from the inert to the harmonious.

Bouyerdene's biography of Abdel Kader, entitled *L'Harmonie des contraires* (*The Harmony of Opposites*), points out that Abdel Kader constantly strove to achieve harmony through balancing the opposites of war and peace, the material and spiritual, East and West, and so on. Abdel Kader is described by his biographer as the incarnation of harmony through his techniques of the self. These techniques include overcoming the fear of combat and suffering, overcoming the desire for vengeance, and being a witness to the unity of being. The anticolonial struggle was a key testing ground for Abdel Kader as he was able to link the courage required for the external struggle or jihad to the interior one. This resembles Gandhi's link between anticolonial struggle and inner transformation.

Nietzsche understood that the basis of bourgeois and modern morality was what he called *ressentiment* and saw modern society as based on repressed desire, repressed violence, and repressed action.[24] Understood from a Yogic point of view, it is a society that seeks to repress the *rajasic* into the *tamasic*, the active into the inert, the violent into the repressed in order to impose a vision of law and equality. This order of law follows inexorably from the state holding the monopoly of violence. The techniques of the transformation of ego, *sattwa*, and the greater or inner jihad have generally in modern bourgeois societies been repressed in favor of a general equality.

The Western binary conception between revolt and harmony does not apply in traditional societies where the metaphysical dominates the material. Revolts and spontaneity are often the product of an ethical transformation or sometimes the revolt of an ethical order under threat from materialism. Order can be simply a structure of repression, but harmony involves techniques of self and the transformation of ego. In the Western colonial regime, order and law were established through repression, producing what I call an apathetic order, often characterized by addiction and consumption. Although the colonized sometimes broke free into passion and violent revolt, in the philosophy of both Gandhi and Abdel Kader, primary struggle must transform its own passions into harmonious states. This process allows development of a type of active passivity.

Charles Taylor explains the transformations of self-fashioning during the sixteenth century in Western Europe.[25] We can argue that during this period, an emotional structure of the European modern individual was encouraged. It involved repressing emotions, desires, anger, and will to power in favor of a semblance of peace and harmony called tolerance, which in fact was simply an effort to impose a type of rationalization of behavior and also obedience to an arbitrary law. Colonialism continued this process outside its borders, combining brutality and the ideology of a civilizing mission. This process was often conceived of as order. The anticolonial struggles in India and Algeria drew upon traditional practices and spiritual philosophies in which the techniques of self defined the primary struggle as being that of the individual against the ego, The individuals and their struggles remain moral examples to us, despite the failure of liberated colonial societies to realize the ideals of behavior established by their founders. Their lives and philosophies and the anticolonial revolts they led demonstrate the links between spontaneity and harmony and challenge the notion that spontaneity and harmony are in opposition.

Endnotes

INTRODUCTION

1. Roger Ames and David Hall, *Democracy of the Dead: Dewey, Confucius, and the Hope for Democracy in China* (Chicago: Open Court, 1999).
2. Friedrich Schlegel, *Kritische Friedrich Schlegel Ausgabe*, ed. E. Behler with J. J. Anstett and Hans Eichner (Paderborn: F. Schoningh, 1958), 18:24, note 64.
3. Karl Marx, *Capital: A Critique of Political Economy*, vol. 1, trans. Ben Fowkes (1867; repr. Harmondsworth: Penguin, 1976), 926.

CHAPTER 1

1. A. N. Whitehead, *Modes of Thought* (New York: Macmillan, 1938). In chapter 3, "Understanding," he argues for the notion of inconsistency as the foundation of logic and for the equally fundamental status of confusion and order. He accuses the great men of history—Epicurus, Plato, Aristotle, and Kant—of committing the fallacy of misplaced concreteness in being unaware of the perils of abstraction in their various claims about certainty. See especially pages 60–63. John Dewey describes the abstracted "rational" sense of order—the selecting out and privileging of one element in the process of experience as causal and foundational—as nothing less than *the* philosophical fallacy; John Dewey, *The Middle Works, 1899–1924*, ed. Jo Ann Boydston (Carbondale: Southern Illinois University Press, 1976–1983), 1:162.
2. I develop some of the ideas in this essay further in Roger Ames, *Confucian Role Ethics: A Vocabulary* (Hong Kong: Chinese University Press; Honolulu: University of Hawaii Press, 2011)
3. John Dewey, *The Essential Dewey*, vol. 1, ed. Larry Hickman and Thomas Alexander (Bloomington: Indiana University Press, 1998), 115.
4. William James, *Pragmatism and Other Writings* (New York: Penguin, 2000), 315.
5. Dewey, *Essential Dewey*, 118.
6. Ibid., 117.
7. Hilary Putnam, *Realism with a Human Face* (Cambridge: Harvard University

Press, 1990), 28. The Confucian position of "embodying experience" (*ti* 體) and "pursuing a ritual propriety in our roles and relations" (*li* 禮) differs from Putnam's position only in that it is more radical, going beyond twentieth-century philosophers' obsession with "language" and "mind" to claim that indeed our entire psycho-physical persons are involved in the process of assimilating and transforming the world as it is experienced. See Ames, *Confucian Role Ethics*, chapter 3. (italics in original).

8. Putnam, *Realism with a Human Face*, 178.
9. Hilary Putnam, *The Many Faces of Realism* (La Salle, IL: Open Court, 1987), 83.
10. Dewey, *Essential Dewey*, 118.
11. George Herbert Mead, *The Individual and the Social Self: Unpublished Work of George Herbert Mead*, ed. David L. Miller (Chicago: University of Chicago Press, 1982), 156.
12. In James's *Pragmatism* (page 41–48), he is explicit in rejecting the notion of some underlying "substance" as a reduplicative fiction that would seem to dispense with the notion of self as such. In other places, however, there is a debate over whether or not James defaults to a more foundational individualism.
13. John Dewey, *Human Nature and Conduct* (New York: Henry Holt and Company, 1922), 176.
14. Ibid., 176–177.
15. Dewey, *Essential Dewey*, 297.
16. John Dewey, *The Early Works, 1892–1898*, ed. Jo Ann Boydston (Carbondale: Southern Illinois University Press, 1969–1972), 3:304.
17. "Dewey: Lectures. Electronic edition," *Intelex Past Masters*, 2013, http://www.nlx.com/collections/147, accessed November 10, 2013.
18. John Dewey, *Reconstruction in Philosophy* (New York: Beacon Press, 1960), 173.
19. See *Analects* 12.1: "Through self-discipline and observing ritual propriety one becomes consummate in one's conduct." 克己復禮為仁. All translations are based upon Roger T. Ames and Henry Rosemont Jr., *The Analects of Confucius: A Philosophical Translation* (New York: Ballantine, 1998).
20. "For Confucius, unless there are at least two human beings, there can be no human beings." Herbert Fingarette, "The Music of Humanity in the *Conversations of Confucius*," *Journal of Chinese Philosophy* 10, no. 4 (1983), 217.
21. Zhu Xi, *Collected Commentaries on the Four Books* (Taipei: Yiwen yinshuguan, 1969): daxue 1a–1b: 大學之道，在明明德，在親民，在止於至善。知止而後有定，定而後能靜，靜而後能安，安而後能慮，慮而後能得。物有本末，事有終始，知所先後，則近道矣。
22. Ibid., daxue 1b–2a: 古之欲明明德於天下者，先治其國；欲治其國者，先齊其家；欲齊其家者，先修其身；欲修其身者，先正其心；欲正其心者，先誠其意；欲誠其意者，先致其知，致 知在格物。物格而後知至，知至而後意誠，意誠而後心正，心正而後身修，身修而後家齊，家齊而後國治，國治而後天下平。

23. Dewey describes such cultivated habits as "habitudes": "The influence of habit is decisive because all distinctively human action has to be learned, and the very heart, blood, and sinews of learning is the creation of habitudes.... Habit does not preclude the use of thought, but it determines the channels within which it operates." Dewey, *Essential Dewey*, 299.
24. Zhu Xi, *Collected Commentaries*, daxue 2b: 自天子以至於庶人，壹是皆以修身為 本。其本亂而末治者否矣，其所厚者薄，而其所薄者厚，未之有也！
25. D.C. Lau and Chen Fong Ching, *A Concordance to the Liji* (Hong Kong: Commercial Press, 1992), 43.1/164/30: 此謂知本，此謂知之至也。
26. Marcel Granet, *La pensée chinoise* (Paris: Editions Albin Michel, 1934), 478. no need for some ultimate One that explains the many.
27. Ames and Rosemont Jr., *Analects of Confucius*.
28. See Ames, *Confucian Role Ethics*, chapter 3, for an extended discussion of the Confucian project.
29. The first dictionary definition of *harmony* is "the act or state of agreeing or conforming," appealing to synonyms such as *accordance, agreement, chime, conformance, conformation, conformity, congruence, congruity, correspondence, harmonization*, and *keeping*.
30. John Dewey, *Art as Experience* (New York: Penguin Group, 1934), 13.
31. Ibid., 16.
32. Ibid., 167.
33. Ames and Rosemont Jr., *Analects of Confucius*, 1.12: 禮之用，和為貴。先王之道斯為美，小大由之。有所不行，知和而和，不以禮節之，亦不可行也。 See also 12.1 and 12.15.
34. Brian Bruya, "The Rehabilitation of Spontaneity," *Philosophy East and West* 60, no. 2 (April 2010), 225.
35. Dewey, *Art as Experience*, 74.
36. Ibid., 291.
37. Ibid.
38. See Zhuangzi, *A Concordance to Chuang Tzu*, Harvard-Yenching Sinological Index Series, Supplement 20 (Peking: Harvard-Yenching, 1947), 7/3/2-5: 庖丁為文惠君解牛，手之所觸，肩之所倚，足之所履，膝之所踦，砉然嚮然，奏刀騞然，莫不中音。合於《桑林》之舞，乃中《經首》之會。文惠君曰："譆！善哉！技蓋至此乎？" 庖丁釋刀對曰：臣之所好者道也，進乎技矣。

39. Ames and Rosemont, *Analects of Confucius*, 2.4: 子曰：吾十有五而志于學，三十而立，四十而不惑，五十而知天命，六十而耳順，七十而從心所欲，不踰矩。

CHAPTER 2

1. This paper is a complete rewrite of a section in chapter 15 of my *Religion, Redemption and Revolution: The New Speech Thinking of Franz Rosenzweig and Eugen Rosenstock-Huessy* (Toronto: University of Toronto Press, 2012).
2. Eugen Rosenstock-Huessy, *The Christian Future or the Modern Mind Outrun* (1946; repr. New York: Harper and Row 1966), 42.
3. Ibid.
4. David Hall and Roger Ames, *The Democracy of the Dead: Dewey, Confucius, and the Hope for Democratic China* (Chicago: Open Court, 1999).
5. Indeed, the liberal versus communitarian debate was very dominant in political theory literature coming out of the United States in the 1980s and early 1990s; Most of those involved in the liberal-communitarian debate simply do not address the deeper issues of history, religion, and trauma that provide the fabric of Rosenstock-Huessy's work.
6. Rosenstock-Huessy, *Christian Future*, 43.
7. Ibid., 47.
8. Ibid., 43: "I suggest that the Theosophical Society has not imported into America one per cent of the Oriental thinking which has been introduced by pragmatism."
9. Ibid., 512.
10. Ibid., 180.
11. Ibid., 48.
12. For the relatively low regard in which he held philosophy, compare this remark (ibid., 11): "[The] Church and Man are in a more crucial situation and … the Cross is more real than theology or philosophy cares to admit."
13. His thesis was developed before the Chinese Revolution, but he rightly saw that the Chinese Revolution unwittingly brought non-European peoples into the revolutionary/ historical logic of the West.
14. *The Collected Works of Eugen Rosenstock-Huessy* (Norwich, VT: Argo, 2005), DVD, vol. 12, lecture 3. Also see the reference to a personal letter by Harold Stahmer in Eugen Rosenstock-Huessy, ed., *Judaism Despite Christianity: The "Letters on Christianity and Judaism" between Eugen Rosenstock-Huessy and Franz Rosenzweig* (1969; repr. New York: Schocken, 1971), 1.
15. Eugen Rosenstock-Huessy, *Out of Revolution* (1938; repr. Norwich, VT: Argo, 1968), 217.

16. Rosenstock-Huessy, *Christian Future*, 10.
17. Ibid., 41.
18. Ibid., 128.
19. Ibid., 61.
20. Ibid., 126–27.
21. Kant, of course, has a darker view of human nature, but he still believes that reason is the way out of that darkness.
22. Rosenstock-Huessy, *Christian Future*, 44.
23. Ibid., 46.
24. Ibid.
25. Ibid.
26. Ibid.
27. Ibid., 51.
28. Ibid., 50.
29. Ibid., 51.
30. Ibid.
31. Bill Ratliff e-mail to author, June 16, 2013.
32. Rosenstock-Huessy, *Christian Future*, 19–20.
33. Harold Berman, *Law and Revolution*, vol. 2, *The Impact of the Protestant Reformations on the Western Legal Tradition* (Cambridge, MA: Harvard University Press, 2003), 20.
34. See chapter 7 of Gary Bullert's *The Politics of John Dewey* (New York: Prometheus, 1983).
35. Rosenstock-Huessy, *Christian Future*, 52.
36. Ibid., 52–53.
37. Ibid., 197.
38. Eugen Rosenstock-Huessy, *Soziologie*, vol. 2, *Die Vollzahl der Zeiten* (Stuttgart: Kohlhammer, 1958), 722.
39. Ibid.
40. Rosenstock-Huessy, *Christian Future*, 178–180; Rosenstock-Huessy, *Soziologie*, 723.

CHAPTER 3

1. I am indebted to my PhD supervisor, Chris Fraser (whose helpful comments and suggestions have greatly improved this chapter), and my MA supervisor, Hsien Chung Lee, for their encouragement and inspiration. Many ideas I have drawn upon here owe a great deal to their published works as well as to ideas and insights expressed in their seminars and other unpublished communications. I hope they will excuse the many omissions

of explicit acknowledgement of their work in this paper.
2. *Mozi* (墨子) is the title of the transmitted collection of Mohist writings; Mozi is the name of the historical person to whom early Mohist teachings are attributed. Mohists were people who identified with the teachings of Mozi and aimed to put them into practice.
3. Angus Graham, *Later Mohist Logic, Ethics and Science* (Hong Kong: Chinese University Press, 2003), 69.
4. Chad Hansen, *A Daoist Theory of Chinese Thought* (New York: Oxford University Press, 1992), 95–98.
5. On this, see Hui-chieh Loy, "Justification and Debate: Thoughts on Moist Moral Epistemology," *Journal of Chinese Philosophy* 35, no. 3 (2008): 457–58.
6. Dan Robins argues that the Mohists were primarily arguing against people they referred to simply as "the gentlemen of today" rather than against the *ru* or early Confucians specifically. Here I will not attempt to deal with the historical details of whether Mohists saw themselves as attacking Confucians, Ruists, or various "gentlemen" or of the changing use of the term *ru*. I merely point out that many of the views the Mohists condemned would later be classified as Confucian in nature and would be closely associated with ideas already present in the Confucian *Analects*. See Dan Robins, "The Mohists and the Gentlemen of the World," *Journal of Chinese Philosophy* 35, no. 3 (2008), 385–402.
7. Though later texts denounced the Mohists for supposedly not endorsing these values, the Mohists themselves appealed to the breakdown of these as exemplary of disorder, and much of the *Mozi* is devoted to arguing how to avoid disorder. See Robins, "Mohists and Gentlemen of the World," 386–88.
8. For example, the *Analects* 8.20. References to the *Analects* here cite section numbers in D. C. Lau and Chen Fong Ching, eds., *A Concordance to the Lunyu* (Hong Kong: Commercial Press, 1995). Locations of all textual references in this paper can also be determined using the Chinese Text Project concordance tool: http://ctext.org/tools/concordance.
9. The *Zhuangzi*, for example, describes how they each "affirm what the other denies, and deny what the other affirms"; *A Concordance to Chuang Tzu*, Harvard-Yenching Institute Sinological Index Series, No. 20 (Repr. Shanghai: Guji, 1986), 4/2/26.
10. The term *li*, variously translated as "rites," "ritual," and "ritual propriety," among others, refers to wide-ranging norms that are seen as both culturally and ethically important by Confucians. It incorporates aspects of ritual (specifying the wearing of particular garments on certain occasions), propriety (rules of social interaction), and rites (stipulations on how to correctly

carry out a funeral and properly conduct mourning). The *li* Confucius refers to in the *Analects* is not his own creation but a collection of existing norms passed down by tradition, to which he attaches great importance.
11. *Analects* 15.18.
12. *Analects* 2.3.
13. I do not mean to suggest that this is the only important aspect of *li* but merely that it is one aspect particularly relevant to this discussion.
14. *Analects* 2.5.
15. *Analects* 8.2.
16. *Analects* 14.41.
17. *Analects* 12.1.
18. For example, *Analects* 9.3.
19. For example, the case described in the *Mengzi* of a man's sister-in-law falling into water and needing to be rescued; though the rites say no one must come into contact with her, other principles override this concern in this case. See D. C. Lau and Chen Fong Ching, eds., *A Concordance to the Mengzi* (Hong Kong: Commercial Press, 1995), 7.17/38/20–30.
20. See Hansen, *Daoist Theory*, 66.
21. *Analects* 1.12.
22. *Mozi* 39/25/75–76. References to the *Mozi* cite page, section, and line numbers given in *A Concordance to Mo Tzu*, Harvard-Yenching Institute Sinological Index Series, No. 21 (Repr. Shanghai: Guji, 1986). Translations are adapted from Y. P. Mei, *The Ethical and Political Works of Motse* (London: Arthur Probsthain, 1929).
23. For example, "宜弟 *yi di* (propitious for younger brothers)," which appears in the *Book of Poetry*, is actually quoted from and appealed to by the Confucian text the *Great Learning* in the context of ordered governance.
24. *Mozi* 3/4/1.
25. In this sense the Mohists probably viewed Confucians as having a standard—the *li*—but as having chosen an incorrect and unjustified standard, as well as failing to follow it consistently.
26. Hansen, *Daoist Theory*, 109.
27. *Mozi* 4/4/9.
28. *Mozi* 45/27/73.
29. *Mozi* 24/16/1.
30. Hsien Chung Lee, *Moxue: Lilun yu Fangfa* (Taipei: Yang-Chih, 2003), 72–80.
31. However, there are also those who argue that the Mohists advocate a divine command theory. For a refutation, see David Soles, "Mo Tzu and the Foundations of Morality," *Journal of Chinese Philosophy* 26, no. 1 (1999): 37–48, and Chris Fraser, "Mohism," *Standard Encyclopedia of Philosophy*, March 23,

2010, *http://plato.stanford.edu/entries/mohism/*, section 7, accessed November 11, 2013.
32. *Mozi* 40/26/9.
33. *Mozi* 41/26/30–36.
34. *Mozi* 11/9/51–59.
35. See Fraser, "Mohism," section 7.
36. *Mozi* 43/27/31–32.
37. For example, the Han Dynasty text *Baihutong* asserts precisely that "people are all born by heaven (人皆天所生也)." The earlier *Lüshi Chunqiu* text states that "heaven gives birth to people (天生人)."
38. See Fraser, "Mohism," section 3.
39. *Mozi* 31/19/15–16.
40. For example, the repeated refrain "I do not know why it is that the scholars of the world hear of inclusivity yet oppose it" after arguments for inclusive care; and "is this not perverse?" after fallacies of opponents are pointed out.
41. For a detailed discussion of the Mohist theory of motivation, see Chris Fraser, "Mohism and Motivation," In Chris Fraser, Dan Robins, Timothy O'Leary, eds., Ethics in Early China. (Hong Kong: Hong Kong University Press).
42. For example, *Analects* 11.18.
43. For example, *Analects* 12.22.
44. *Mozi* 40/26/1, 45/28/1–2, 89/49/24–27.
45. *Mozi* 28/17/7–29/17/14.
46. *Analects* 13.4.
47. Mengzi 1.7/4/20.
48. *Analects* 8.9.
49. *Analects* 5.16.

CHAPTER 4

1. The argument of the complementarity principle in atomic physics was advanced by Niels Bohr. He introduced it for the first time in public in 1927 in Como, Italy, in the lecture "The Quantum Postulate and the Recent Development in Atomic Theory" at the International Physical Congress; cf. Niels Bohr, *The Philosophical Writings of Niels Bohr*, vol. 2, *Essays 1932–1957 on Atomic Physics and Human Knowledge* (Woodbridge, CT: Ox Bow Press, 1971). According to Heisenberg, Bohr developed his ideas on complementarity while on vacation in Norway; cf. Max Jammer, *The Conceptual Development of Quantum Mechanics* (New York: McGraw-Hill Book Company, 1966), 347–51. Simply put, the principle states that an exhaustive description of quantum

phenomena is possible only through recourse to two mutually exclusive sets of classical concepts. One cannot employ "either/or" or "both/and" with Bohr's complementarity principle since the former implies that one of the explanations is enough and the latter implies that both can apply simultaneously. The author's "mixed salad" description of "either/and" was greeted with a positive response by Finn Aaserud, director of the Niels Bohr Archives, when it was put forth in a private conversation at the archives on October 20, 1995. Heisenberg's uncertainty principle is a special case of the complementarity principle. (This would favor the epistemological interpretation of the uncertainty principle.) Cf. Sandro Petruccioli, *Atoms, Metaphors and Paradoxes: Niels Bohr and the Construction of a New Physics* (Cambridge: Cambridge University Press, 1993). (In the complementarity principle advanced in this work, it is not suggested that the two different points of view [Hegelian and Taoist] are mutually exclusive.)

2. Hanna Rosental was the wife of physicist Stefan Rosental, who was Bohr's right-hand man for many years. Husband and wife escaped together from Nazi Germany. Hanna Rosental, a historian, was a former classmate of Hanna Arendt's and a pupil of Edmund Husserl's. In fact, she and Arendt were in Husserl's class together. I was pleased to spend a fine evening with her at Finn's home, lit throughout with candles in the Danish Christmas tradition, and was regaled with Hanna's stories of how their boat was nearly apprehended by Nazi surveillance during their escape from Denmark to Sweden, along with revealing personal reminiscences of Husserl and rapier-sharp observations of her classmate Arendt.

3. Compare the argument in Robert E. Allinson, "Moral Values and the Taoist Sage in the *Tao de Ching*," *Asian Philosophy: An International Journal of the Philosophical Traditions of the East* 4, no. 2 (1994), 127–36.

4. Rasmussen points out: "In the very first sentence of his 1958 contribution to Kilbansky's *Philosophy of the Mid-Century* Bohr said, 'The significance of physical science for philosophy does not merely lie in the steady increase of our experience of inanimate matter, but above all in the opportunity of testing the foundation and scope of some of our most elementary concepts.'" *Cf*, Erik Rasmussen, *An Essay on Fundamentals of Political Science Theory and Research Strategy* (Odense: Odense University, 1987), 124.

5. *Cf*, Henry J. Folse, *The Philosophy of Niels Bohr: The Framework of Complementarity* (Amsterdam: Elsevier Science Publisher, 1985), 54.

6. *Cf*, Wing-tsit Chan, "The Story of Chinese Philosophy," in *The Chinese Mind: Essentials of Chinese Philosophy and Culture*, ed. Charles A. Moore (Honolulu: East-West Center Press/University of Hawaii Press, 1967), 35.

7. The idea originated with Fichte rather than Hegel, but Hegel borrowed it

and made it famous, and it has thereby been associated with his name.
8. Robert E. Allinson, "An Overview of the Chinese Mind," in *Understanding the Chinese Mind: The Philosophical Roots*, ed. Robert E. Allinson, 10th ed. (New York: Oxford University Press, 2000), 23.
9. This notion of dialectic was already prefigured in *Phaedo*, when Socrates explains how delighted he was when he heard someone reading from a book—which the reader said was written by Anaxagoras—that everything was to be explained by the mind, though Anaxagoras did not make any use of his concept of mind. Socrates subsequently appropriated the concept of mind and extended it to explain why he would stay in prison to accept the penalty of Athens.
10. Samuel P. Huntington, "The Clash of Civilizations?" *Foreign Affairs* 72, no. 3 (summer 1993), 22–49.
11. *Analects*, 3.3.
12. This is said with no relationship to the idea that music that belongs to a corrupt era can cause corruption in the minds of those who enjoy it in another era, which appears to be expressed in Isaiah Berlin's statement, in his account of Tolstoy, that "we find the works of Mozart and Chopin beautiful only because Mozart and Chopin were themselves children of our decadent culture, and therefore their works speak to our diseased minds; but what right have we to infect others, to make them as corrupt as ourselves." *Cf*, Isaiah Berlin, "Tolstoi and Enlightenment," in *Russian Thinkers*, ed. Henry Hardy and Aileen Kelly (Harmondsworth: Penguin Books, 1979), 255. It is entirely possible, though it is not clear from the essay, that Berlin, normally a most perspicacious thinker, was merely restating Tolstoy's view.
13. Hsun-Tzu once spoke of *cheng ming*, or the rectification of names. There is a need now, in the West (or more properly in a global region in which philosophy is to be carried out), to become aware of this essential philosophical activity of the rectification of names; cf. Hsun-Tzu, Book 22. To rectify a name, for Confucianism, was to return the name to its proper use, which meant accordance with its ancient meaning. Of course, this activity meant more than simply using language correctly, as its major emphasis was on orienting persons to engage in appropriate activities. Cf. *Analects*, 13.3.
14. Ibid., 12.5.

CHAPTER 5

1. "*The Analects*," *Internet Classics Archive*, 1994–2009, http://classics.mit.edu/Confucius/analects.html, accessed August 10, 2010.
2. Arthur Waley, trans., *The Analects of Confucius* (1938; repr. New York: Vin-

tage, 1989); Edward Slingerland, "Why Philosophy Is Not 'Extra' in Understanding the Analects," *Philosophy East and West* 50 (2000): 137–41; Edward Slingerland, "Reply to Bruce Brooks and A. Takeo Brooks," *Philosophy East and West* 50 (2000): 146–47.

3. *Analects*, 12.19.
4. Ibid., 2.3; see also 13.6.
5. "Classical Confucian Texts: *Mencius*," *Chinese Philosophical Etext Archive*, October 10, 2005, http://sangle.web.wesleyan.edu/etext/pre-qin/pre-qin.html, accessed August 10, 2010.
6. *Analects*, 12.11.
7. See, for example, Bernard Williams, "Persons, Character, and Morality," in *Moral Luck* (Cambridge: Cambridge University Press, 1981).
8. John Rawls, *A Theory of Justice*, revised ed. (Cambridge, MA: Belknap Press of Harvard University Press, 1999), 3.
9. Ibid., sections 4 and 24.
10. Ibid., section 11.
11. Ibid., sections 5, 27, and 28.
12. "Classical Confucian Texts: *Mencius*," 1B7.
13. "Classical Confucian Texts: *Xunzi*," *Chinese Philosophical Etext Archive*, October 10, 2005, http://sangle.web.wesleyan.edu/etext/pre-qin/pre-qin.html, accessed August 10, 2010.
14. David Wong, "Rights and Community in Confucianism," in *Confucian Ethics: A Comparative Study of Self, Autonomy, and Community*, eds. Kwong-loi Shun and David B. Wong (New York: Cambridge University Press, 2004).
15. Joseph Chan, "A Confucian Perspective on Human Rights for Contemporary China," in *The East Asian Challenge for Human Rights*, ed. Joanne R. Bauer and Daniel A. Bell (Cambridge: Cambridge University Press, 1999).
16. Henry Rosemont, *A Chinese Mirror: Moral Reflections on Political Economy and Society* (La Salle, IL: Open Court, 1991); Henry Rosemont, "Whose Democracy? Which Rights? A Confucian Critique of Modern Western Liberalism," in *Confucian Ethics: A Comparative Study of Self, Autonomy, and Community*, eds. Kwong-loi Shun and David B. Wong (New York: Cambridge University Press, 2004).
17. For example, Henry Rosemont, "Kierkegaard and Confucius: On Finding the Way," *Philosophy East and West* 36, no. 3 (1986): 201–12.
18. Isaiah Berlin, "Two Concepts of Liberty," in *Four Essays on Liberty* (London: Oxford University Press, 1969), 118–72.
19. John Dewey, "The Ethics of Democracy," in *The Early Works 1882–1898*, vol. 1, ed. Jo Ann Boydston (Carbondale: Southern Illinois University Press, 1969–1975); John Dewey, "Christianity and Democracy," in *The Early Works*

1882–1898, vol. 4, ed. Jo Ann Boydston (Carbondale: Southern Illinois University Press, 1969–1975).
20. Dewey, "Ethics of Democracy," 231–32.
21. John Dewey, "The Public and Its Problems," in *The Later Works*, vol. 2, ed. Jo Ann Boydston (Carbondale: Southern Illinois University Press, 1981–1990), 364.
22. John Dewey, "Liberalism and Social Action," in *The Later Works*, vol. 2, ed. Jo Ann Boydston (Carbondale: Southern Illinois University Press, 1981–1990), 56.
23. Dewey, "The Public and Its Problems," 327–28.

CHAPTER 6

1. It is possible that Eastern thought does not conceptualize the contrast between harmony and order and spontaneity and revolt as one of stark opposition as I just did. However, being from the West, trained in Western philosophy, and largely ignorant (hopefully not for long) of the Eastern tradition, I will offer an analysis that is marked by the kind of dualisms and antagonisms of the Western tradition. Since I cannot ambition to develop a comparative perspective with Eastern thought, I will be content in this paper to offer a broad perspective from within Western thought on the way I see this conceptual opposition as having evolved, and to a degree matured, from the ancients' harmonious but static and illiberal political order to the modern defense of complex, self-organizing, and often disharmonious orders.
2. Rebels to the existing order were sent to death, exiled, or socially ostracized. Revolt was to be nipped in the bud as a threat to the entire body politic. The only form of social criticism that was tolerated to a degree was expressed through the work of artists, who were norm breakers by excellence and were pursuers of individual self-expression. In Plato's *Republic*, artists unwilling to abide by his strict regulations on music and poetry were to be banned.
3. This text was written prior to the events of January 2011 that led to regime changes in Tunisia and Egypt. It would be interesting to consider whether what is said here of the East generally is not similarly applicable to what is now left of the Middle Eastern and North African illiberal political orders and whether it may explain why the Egyptian and Tunisian dictatorships were incapable of turning into positive and internally transformative forces their youth's spontaneity and feelings of revolt.
4. I will have little to say, however, about how this Western conceptual achievement may or may not contrast with treatment of the same conceptual tension in Eastern philosophy, since I know virtually nothing about the latter.

5. There are of course various nuances within that ancient view of political order. Plato's own disciple, Aristotle, though sharing the holistic views of his master, defends the notion of a nonrationalistic, organic political order. In *Politics*, Aristotle describes the social order as "by nature" rather than by the design of highly rational minds. Accordingly, man is a "political animal," which means that just as a bee does not exist without the hive or the ant does not make sense without the colony, a human being is not a human being outside the polity. Each individual—man, woman, child, or slave—is consequently primarily defined by his or her function in the social whole, which is also given by his or her telos—that is, the deterministic purpose of an individual's life: the free man rules; women, children, and slaves by nature obey, in proportion to their respective, lesser share of reason. This natural order, however, does not accommodate individual freedom in the modern sense (a telos is a destiny, not freedom). For that, one will need to turn to later, liberal advocates of freedom as a right to pursue your own, deliberately chosen definition of the good.
6. After the article "Spontaneous," *Merriam Webster Online Dictionary*, 2010, http://www.merriam-webster.com/dictionary/spontaneous, accessed August 2010.
7. This fact may say something about the nature of Western philosophy and its tendency to reify dualisms into threatening entities (right and wrong, true and false, just and unjust, and so on).
8. "Order," *The Free Dictionary*, 2013, http://www.thefreedictionary.com/order, accessed August 2010. See also "Order," *Merriam-Webster*, 2013, http://www.merriam-webster.com/dictionary/order, accessed August 2010.
9. For examples from the East, one can think of the difficulty for the Chinese government in dealing with even peaceful manifestations of disagreement, such as the Tiananmen Square tragedy (or "incident," as the Honk Kong Museum of History calls it) or the recent wave of suicides among factory workers at Foxconn Technology, showing the slowness and rigidity of certain hierarchies in recognizing and accommodating social needs. See David Barboza, "After Spate of Suicides, Technology Firm in China Raises Workers' Salaries," *New York Times*, June 2, 2010, http://www.nytimes.com/2010/06/03/business/global/03foxconn.html, accessed November 11, 2013.
10. The only rights they do not surrender are some protohuman rights, including a "right to govern their own bodies; enjoy aire, water, motion, waies to go from place to place; and all things else without which a man cannot live, or not live well"; Thomas Hobbes, *Leviathan* (1651; repr. Indianapolis: Hackett Publishing Company, 1994), 95.

11. "If there is no law made, [the sovereign can punish] according as he shall judge most to conduce to the encouraging of men to serve the Common-wealth, or deterring of them from doing dis-service to the same"; Hobbes, *Leviathan*, 110.
12. Ibid., 115. For Hobbes, this "great inconvenience"—that is, the danger of being deprived of one's possession for the enriching of a flatterer—exists whether power is in the hands of a single man or an assembly.
13. Every man thus has a right to disobey orders that require him to kill or maim, or that surrender him to a certain death, and thus has a right to flee from combat or executioners when imminent death threatens; Hobbes, *Leviathan*, 131–32.
14. John Locke, *Second Treatise of Civil Government* (1690; repr. Indianapolis: Hackett Publishing Company, 1980), 12.
15. Ibid., 74.
16. Ibid., 111.
17. Ibid., 113.
18. Jean-Jacques Rousseau, "On the Social Contract," in *The Basic Political Writings* (Indianapolis: Hackett Publishing Company, 1987), 150.
19. See Isaiah Berlin, "Two Concepts of Liberty," in *Four Essays on Liberty* (London, Oxford University Press, 1969), 118–72.
20. See B. Constant, "On the Liberty of the Ancients Compared with That of the Moderns," in *The Political Writings of Benjamin Constant*, ed. Biancamaria Fontana (Cambridge, MA: Cambridge University Press, 1988), 309–28.
21. See, for example, James Bohman and William Rehg, *Deliberative Democracy: Essays on Reason and Politics* (Cambridge, MA: MIT Press, 1997) or Jon Elster, *Deliberative Democracy* (Cambridge, MA: Cambridge University Press, 1998).
22. John Stuart Mill, *On Liberty and Other Writings* (Cambridge: Cambridge University Press, 1989), 13.
23. John Stuart Mill, *On Liberty* (New York: Longmans, Green, and Company, 1913), 20.
24. What is striking in this passage, though, is a celebration of dissent and mistakes in a way that was not possible with previous thinkers. In Hobbes, dissent was not an option, as the right to tell right from wrong had been surrendered to the Leviathan. In Rousseau, mistakes got canceled out in the calculus of the general will and were certainly not constitutive of it. Only in Locke do we find a celebration of tolerance (in his "Letter on Toleration") that accommodates dissent and heresy, to a degree. Mill, however, goes much further than Locke in saying that not only should society tolerate diverging viewpoints, as one tolerates a necessary evil, but it should treat these dissenting opinions as equally worthy of expression as the majoritarian

ones and as contributing equally to the social pursuit of truth and therefore to social utility.

25. I focus here on *The Public and Its Problems*, which contains clear assertions regarding the properties of public deliberation as a form of social inquiry and the existence of "social intelligence." *Democracy and Education* is also relevant in that it makes clear that democracy is not just a set of political institutions but also a way of life, rooted in a certain type of education. For harmony and social order to emerge, spontaneity must be cultivated at all levels of society.

26. Walter Lippmann, *The Phantom Public* (1925; repr. New Brunswick, NJ: Transaction Publishers, 1993).

27. John Dewey, *The Public and Its Problems* (1927; repr. Chicago: Swallow Press, 1954).

28. Dewey, *Public and Its Problems*, 206–207.

29. Ibid., 207–208.

30. Ibid., 20.

31. "Majority rule, just as majority rule, is as foolish as its critics charge it with being. But it never is merely majority rule" (Dewey, *Public and Its Problems*, 207). What makes it valuable are, among other things, "antecedent debates, modification of views to meet the opinions of minorities, and the relative satisfaction given the latter by the fact that it has had a chance and that next time it may be successful in becoming a majority" (207–208).

32. For Dewey, indeed, a community, including a political community, "must always remain a matter of face-to-face intercourse." Another crucial feature of democratic deliberation for Dewey, which derives directly from this face-to-face dimension, is that deliberation must ultimately be oral rather than written. He writes, "The winged words of conversation in immediate intercourse have a vital import lacking in the fixed and frozen words of written speech" (Dewey, *Public and Its Problems*, 218). Thus while social inquiry and the dissemination of its results can take place in print, "their final actuality is accomplished in face-to-face relationships by means of direct give and take."

33. Habermas actually claims Dewey as an important influence on his own thinking, so it is not far-fetched to see Dewey as a forerunner of his ideas.

34. Dewey, *Public and Its Problems*, 178.

35. Ibid., 184.

36. Hayek is sometimes caricatured as a libertarian and a "free marketeer," committed to the view of minimal government and the elimination of all government interference in the economic sphere. It is true that in some respects Hayek's views about the market were fairly radical. He thought the science of economics should be renamed catallaxy—or the science of exchange—to

emphasize that one should not presuppose shared ends of the people transacting on the market, as the word *economy* does. I will use Hayek's example of the market to illustrate the case of a spontaneous order, but it should be said first that this is not the only type of order Hayek considered. Hayek thought that many norms and the whole tradition of common law were forms of spontaneous order to be contrasted with the rigid and rationalistic order imposed by political will. He thus contrasted the cosmos of the law (sometimes written down as positive law) and the taxis of legislation, which could violate existing laws. Second, far from being an absolutist of cosmos, Hayek does not systematically reject all forms of planned orders or taxis. He thus writes in several places that he believes in an important role for government and considers necessary democratic debates and decisions on a number of issues, including some related to market regulations.

37. Friedrich Hayek, *Law, Legislation, and Liberty*, vol. 1 (Chicago: University of Chicago Press, 1983), 36.
38. This is a reading by James Buchanan: "Order defined in the process of its emergence" is offered to counter the perception that economic order is simply the product of an omniscient designing mind. For Buchanan, Hayek's insight is that even assuming full information at any given time, the cosmos of the market, or any other cosmos, could not be intentionally designed because it exists only as the emergent phenomenon of certain types of individual relations. See "Reader's Forum on Norman Barry's 'The Tradition of Spontaneous Order,'" *Online Library of Liberty*, 2013, http://oll.libertyfund.org/?option=com_content&task=view&id=163&itemid=282, accessed November 11, 2013.
39. Every economic transaction between individuals at any given time contains relevant information that is immediately aggregated in market prices. See F. A. Hayek, "The Use of Knowledge in Society," *American Economic Review* 35, no. 4 (September 1945): 519–30.

CHAPTER 7

1. Enchi Terutake, *Gendai nihon shi shi* (Tokyo: Shoshinsha, 1958), 110.
2. Kitamura Tokoku, "Tokugawa jidai-no heiminteki riso," in *Gendai bungakuron taikei*, vol. 1 (Tokyo: Kawadeshobo, 1954), 248.
3. P. B. Shelley, *Letters, Articles, Excerpts*, vol. 6 (Moscow: Nauka, 1972), 440.
4. R. F. Yusufov, *Russian Romanticism of the Early Nineteenth Century and National Cultures* (Moscow: Nauka, 1970), 72.
5. I. G. Neupokoeva, *Common Features in the European Romanticism and Specifications of the National Schools* (Moscow: Nauka, 1973), 30.

6. Y. N. Nersesov, *Literary Theory of the German Romanticism* (Leningrad: Izdatel'stvo pisateley v Leningrade, 1934), 217.
7. Kitamura Tokoku shu, *Nihon gendai bungaku zenshu*, vol. 9 (Tokyo: Kaizosha, 1927), 231.
8. Quoted in Terutake, *Gendai nihon shi shi*, 122.
9. N. I. Konrad, *Japanese Literature* (Moscow: Nauka, 1974), 293.
10. Tokoku shu, *Nihon gendai bungaku zenshu*, 229.
11. Ibid., 228.
12. Ito Shinkichi, *Nihon no shika*, vol. 26 (Tokyo: Chuokoronsha, 1971), 12.
13. See Shelley, *Letters, Articles, Excerpts*, 434.
14. Tokoku shu, *Nihon gendai bungaku zenshu*, 241.
15. D. T. Suzuki, *Essays in Zen Buddhism*, vol. 1 (London: Luzac, 1934), 18.
16. F. de La Bart, *Research in the Field of Romanticist Poetics and Style* (Kharkov, Ukraine: Silberberg and Sons, 1911), 431.
17. T. P. Grigoryeva, *Japanese Aesthetic Tradition* (Moscow: Nauka, 1979), 136.
18. See L. E. Cherkassky, *New Chinese Poetry (Tradition* (Moscow: Nauka, 1979), 136.
19. See M. F. Ovsyannikov, ed., *History of Aesthetics: The Monuments of the World Aesthetic Thought in Six Volumes*, vol. 3, *Aesthetic Doctrines of Western Europe and the USA (1789–1871)* (Moscow: Iskusstvo, 1967), 793.
20. Quoted in Hirosue Tamotsu, *Zen kindai no kanosei* (Tokyo: Kage Shobo, 1960), 44.
21. Tokoku shu, *Nihon gendai bungaku zenshu*, 142.
22. See Hisamatsu Senichi, *Nihon bungaku shi*, vol. 6 (Tokyo: Shibundo, 1961), 102.
23. R. H. Blyth, *Zen and Zen Classics*, vol. 5 (Tokyo: Hokuseido Press, 1960), 101.
24. Sakamoto Hiroshi, *Kitamura Tokoku* (Tokyo: Shibundo, 1957), 190.
25. Yoshida Seiichi, *Kindai nihon bungaku chi: Meiji-Taisho hen* (Tokyo: Yamada Shoten, 1957), 75.
26. Ino Kenji, *Kindai nihon-no bungaku* (Tokyo: Fukumura shoten, 1954), 44.
27. Ishimaru Hisashi, "Wakanashu kara rakubaishu e," in *Shimazaki Toson kenkyu* (Tokyo: Yuseido Shuppan, 1961), 44.
28. Togawa Shukotsu, "Kinnen no bunkai ni okeru ancho," in *Gendai bungakuron taikei* (Tokyo: Kawadeshobo, 1954), 125.
29. Ito Shinkichi, *Nihon no shika*, 14.
30. Tokoku shu, *Nihon gendai bungaku zenshu*, 232.

CHAPTER 8

1. Denis Diderot, *Rameau's Nephew and D'Alembert's Dream*, trans. Leonard Tan-

cock (Harmondsworth: Penguin, 1966), 97.
2. Ibid., 96.
3. Ibid., 93.
4. Ibid., 96.
5. Ibid., 93.
6. Ibid., 96.
7. Thomas De Quincey, "On Murder Considered as One of the Fine Arts," in *The Works of Thomas De Quincey*, 21 vols., ed. Grevel Lindop (London: Pickering and Chatto, 2000–2003), 6:112.
8. Ibid., 6:113.
9. Ibid., 6:124–25. In the essay, De Quincey toys with the idea that "every philosopher of eminence for the two last centuries"—that is, Descartes, Spinoza, Hobbes, and Malebranche, in addition to Kant—"has either been murdered, or, at the least, been very near it; insomuch, that if a man calls himself a philosopher, and never had his life attempted, rest assured there is nothing in him" (118). Thus De Quincey considers it "an unanswerable objection (if we need any)" against Locke's philosophy that "although he carried his throat about with him in this world for seventy-two years, no man ever condescended to cut it" (118). On the other hand, Leibniz must have felt deeply "insulted by the security in which he passed his days." His ambition that there would at least be an attempt on his life—which would bring ultimate recognition of his philosophy—was so great that in his old age he amassed a large sum of gold and kept it in his house to attract a potential murderer, but without success. In the end, he died "partly of the fear that he should be murdered, and partly of vexation that he was not" (124).
10. Ibid., 6:131.
11. Ibid., 6:132–33.
12. De Quincey, "Second Paper on Murder Considered as One of the Fine Arts," in *The Works of Thomas De Quincey*, 21 vols., ed. Grevel Lindop (London: Pickering and Chatto, 2000–2003), 11:405.
13. De Quincey, "On Murder," 6:125–26.
14. Letter to Sophie Volland, September 30, 1760, in Denis Diderot, *Diderot's Letters to Sophie Volland*, trans. Peter France (London: Oxford University Press, 1972), 61.
15. Ibid. This motif is quite frequent in seventeenth- and eighteenth-century philosophy; see, for example, Pierre Bayle, *Various Thoughts on the Occasion of a Comet*, trans. Robert C. Bartlett (Albany: State University of New York, 2000), 200: "Despite the wheel, the zeal of the magistrates, and the diligence of the provosts, how many murders and how much brigandage are carried out in the very place and time that criminals are executed?" See also

Paul-Henri Thiry d'Holbach, "Système de la Nature," in *Œuvres philosophiques complètes*, 7 vols., ed. Jean-Pierre Jackson (Paris: Editions Alive, 1999), 2:305, note 19: "On vole tous les jours au pied même des échafauds où l'on punit les coupables"; and Paul-Henri Thiry d'Holbach, "Éthocratie," in *Œuvres philosophiques complètes*, 7 vols., ed. Jean-Pierre Jackson (Paris: Editions Alive, 1999), 3:684, note 87: "Chacun sait qu'il se commet presque toujours des filouteries et des vols aux pieds des échafauds et des potences où la peuple en foule va voir les executions."

16. Saint Augustine, *The Confessions of Saint Augustine*, trans. Edward Bouverie Pusey (Rockville, MD: Arc Manor, 2008), 24–28.
17. Letter to Paul Landois, June 29, 1756, in Denis Diderot, *Œuvres*, 5 vols., ed. Laurent Versini (Paris: Robert Laffont, 1994–1997), 5:56.
18. De Quincey, "On Murder," 6:115.
19. Letter to Sophie Volland, October 14–15, 1760, in Diderot, *Œuvres*, 5:255.
20. Denis Diderot, *Jacques the Fatalist*, trans. Michael Henry (Harmondsworth: Penguin, 1986), 25.
21. Letter to Paul Landois, June 29, 1756, in Diderot, *Œuvres*, 5:56.
22. Diderot, *Rameau's Nephew*, 218.
23. Diderot, *Encyclopédie* s.v. "Droit naturel," *Œuvres*, 3:44.
24. Dennis Diderot, *Jacques the Fatalist* (Oxford: Oxford University Press, 2008), 78–79.
25. Denis Diderot, "Letter on the Blind for the Use of Those Who See," in *Thoughts on the Interpretation of Nature and Other Philosophical Works*, ed. David Adams, trans. Margaret Jourdain (Manchester: Clinamen Press, 1999), 155.
26. Letter to Paul Landois, June 29, 1756, Diderot, *Œuvres*, 5:56.
27. Diderot, *Rameau's Nephew*, 216.
28. Letter to Sophie Volland, September 30, 1760, Diderot, *Letters to Sophie Volland*, 61.
29. Diderot, *Rameau's Nephew*, 108.
30. Denis Diderot, "Salon de 1765," *Œuvres*, 4:380.
31. Letter to Sophie Volland, September 30, 1760, Diderot, *Letters to Sophie Volland*, 61.
32. Letter to Sophie Volland, October 14–15, 1760, Diderot, *Œuvres*, 5:255.
33. For a detailed account of Diderot's philosophy of energy, see Jacques Chouillet, *Diderot, poète de l'énergie* (Paris: Presses Universitaires de France, 1984).
34. G. W. Leibniz, *Theodicy*, trans. E. M. Huggard (La Salle: Open Court, 1985), 373.
35. Leibniz, "Discourse on Metaphysics," in *Philosophical Writings*, trans. Mary Morris and G. H. R. Parkinson (London: Dent, 1973), 40.

36. Letter to Sophie Volland, October 14–15, 1760, Diderot, *Œuvres*, 5:254.
37. Ibid., 254–55.
38. See Sigmund Freud, *Jokes and Their Relation to the Unconscious*, trans. James Strachey (London: Vintage, 2001), 229. As an example of gallows humor, Freud cites a criminal being led to execution on a Monday who remarks, "Well, this week's beginning nicely," and another criminal, being led to execution on a cold day, who asks for a scarf in order not to catch cold. The latter criminal's concern for his health is superfluous in view of the fate that awaits him, just as the former criminal's remark is "misplaced … since for the man himself there would be no further events that week." Damiens's complaint is similarly misplaced; he complains about the hard day ahead of him as though he were thinking of sparing his strength for something he needed to attend to that evening. While Freud admires the two criminals for their tenacious hold upon their customary selves, even in the most extraordinary of circumstances, Diderot admires not only that Damiens remained true to his nature to the very end but also that nature itself, even if it is "atrocious."
39. Diderot, "Salon de 1765," *Œuvres*, 4:380.
40. Diderot, *Rameau's Nephew*, 89.
41. Ibid., 93.

CHAPTER 9

1. René Guénon, *East and West* (Hillsdale, NY: Sophia Perennis, 2002).
2. Seyyed Hossein Nasr, *Man and Nature: The Spiritual Crisis in Modern Man* (New York: Harper Collins, 1991).
3. M. K. Gandhi, "Statement to Disorders Inquiry Committee, January 5, 1920," in *The Collected Works of Mahatma Gandhi*, vol. 19 (New Delhi: Publications Division Ministry of Information and Broadcasting, Government of India, 2011), 206.
4. Paul Deussen, *The Philosophy of the Upanishads* (New York: Cosimo Books, 2006).
5. See I. K. Taimni, *The Science of Yoga: The Yoga-Sutras of Patanjali in Sanskrit* (Madras: Theosophical Publishing House, 1979).
6. R. K. Prabhu and U. R. Rao, "Between Cowardice and Violence," in *The Mind of Mahatma Gandhi* (Ahmedbad: Navajivan Publishing House, 1941).
7. Mahatma Gandhi, *All Men are Brothers* (New York: Continuum Publishing, 2005), 97.
8. Eknath Easwaran, *Gandhi, the Man: The Story of His Transformation* (Tomales, CA: Nilgiri Press, 1997),14.9.

9. Easwaran, *Gandhi*, 35.
10. Ibid.
11. "The Trial of Mahatma Gandhi—1922," *High Court of Bombay*, 2013, http://bombayhighcourt.nic.in/libweb/historicalcases/cases/TRIAL_OF__MAHATMA_GANDHI-1922.html, accessed January 16, 2011.
12. Ibid.
13. See Taimni, *Science of Yoga*.
14. E. Levinas, *Totality and Infinity*, trans. Alphonso Lingis (Pittsburgh: Duquesne University Press, 1969), 238–39.
15. For example for this idea of jihad as inherently warlike, see Andrew C. McCarthy, *The Grand Jihad: How Islam and the Left Sabotage America* (New York: Encounter Books, 2010).
16. Michel Chodkiewicz, *The Spiritual Writings of Amir 'Abd al-Kader* (Albany: State University of New York Press, 1995).
17. Antoine-Adolphe Dupuch, *Abd El-Kader au Chateau d'Amboise* (Paris: Ibis Presse, 2002), 21–22.
18. Reza Shah-Kazemi, "From the Spirituality of Jihad to the Ideology of Jihadism," in *Islam, Fundamentalism, and the Betrayal of Tradition: Essays by Western Muslim Scholars*, ed. Joseph E. B. Lumbard (Bloomington, IN: World Wisdom, 2004),130.
19. F. Skali, *Futuwah: Traité de chevalerie soufie* (Paris: Albin Michel), 1989.
20. Michael Leicht, *M. Gandhi: His Philosophical and Religious Thought and Some Cross References* (Nordestedt: Grin Verlag Hausarbeit, 2006), 18.
21. Reza Shah-Kazemi, "From the Spirituality of Jihad," 130.
22. Leon Roche, who pretended to convert to Islam and to be at the emir's service in order to spy on him, related the following incident: "I was admitted some times to his tent where I had the honor to sleep I saw him in prayer and I was struck by his mystical states but this night he gave a powerful image of faith which must have been how the great saints of Christianity appeared"; A. Bouyerdene, *Abd El-Kader: L'harmonie des contraires* (Paris: Seuil, 2008), 10 (my translation).
23. Ndiouga Kebe, "La formation spirituelle du *murīd* selon Abū Ṭālib al-Makkī," in *Les maîtres soufis et leurs disciples des IIIe-Ve siècles de l'hégire (IXe-XIe): Enseignement, formation et transmission*, eds. Geneviève Gobillot and Jean-Jacques Thibon (Beirut: Presses de L'Ifpo, 2012).
24. F. Nietzsche, *The Genealogy of Morals*, trans. Horace Barnett Samuel (New York: Courier Dover Publications, 2003).
25. Charles Taylor, *A Secular Age* (Cambridge, MA: Harvard University Press, 2007).

Index

A

Abu Talib 167
Adolphus 145, 146
al Maki 167
Aristotle 14, 15, 47, 76, 90, 102

B

Bach 105
Balthasar 41
Basho 125, 128, 133
Benjamin 10, 102, 184
Bentham 110, 148, 197
Berlin 15, 109, 110, 180, 181, 184
Berman 13, 45, 175
Bohr 10, 67, 68, 69, 70, 71, 178, 179
Bo Juyi 133
Bopichon 166
Buddha 49, 50

C

Cato of Utica 151
Chikamatsu 134
Chrysippus 32
Collingwood 75
Confucius 5, 6, 7, 9, 10, 12, 24, 28, 29, 34, 36, 37, 41, 42, 43, 44, 47, 49, 51, 55, 59, 63, 64, 69, 71, 83, 86, 87, 89, 93, 171, 172, 173, 174, 177, 180, 181, 197

D

Dante 83
De Quincey 144, 145, 146, 149, 152, 188, 189
Descartes 8, 9, 156, 188
Dewey 5, 6, 9, 10, 19, 20, 21, 22, 23, 24, 29, 30, 31, 33, 34, 35, 36, 37, 41, 42, 43, 44, 47, 48, 49, 51, 97, 98, 99, 104, 110, 112, 113, 114, 117, 171, 172, 173, 174, 175, 181, 182, 185, 197, 199
Diderot 11, 141, 142, 145, 146, 147, 148, 149, 150, 151, 152, 153, 187, 188, 189, 190
Doi Bansui 127
Dostoevsky 127
Du Fu 133

E

Edgar Allan 130
Emir Abdel 11, 155, 165
Epicurus 32, 171

F

Friedrich 11, 17, 104, 110, 112, 115, 171, 186
Futabatei Shimei 127, 128

G

Gandhi 11, 155, 156, 157, 158, 159, 160, 161, 162, 163,

164, 165, 166, 167, 168, 190, 191
George Herbert 22, 172
Goethe 42, 141
Granet 27, 173
Guénon 155, 190

H

Habermas 113, 185
Hartmann 124, 128, 129
Hayek 17, 99, 104, 106, 110, 112, 113, 115, 116, 117, 185, 186
Hegel 10, 67, 72, 76, 164, 179
Hisamatsu Senichi 135, 187
Hobbes 97, 103, 107, 108, 109, 110, 183, 184, 188

I

Ikkyu 133
Ishizaka Minako 134

J

James 20, 21, 22, 34, 103, 171, 172, 184, 186, 190
James Carse 34
Jesus 163
Ji Kangzi 87
Joachim of Flore 40
John Stuart 104, 110, 117, 184

K

Kader 11, 155, 156, 165, 166, 167, 168, 191
Kant 13, 70, 90, 144, 171, 175, 188
Keats 132, 136
Kierkegaard 9, 70, 181
Kitamura Tokoku 122, 126, 136, 186, 187
Kunikido Doppo 132

L

Lao Tzu 5, 6, 9, 10, 35, 50, 51, 67, 68, 70, 82
Lermontov 136
Levinas 10, 164, 165, 191
Li Po 125, 133
Lippmann 113, 185
Locke 42, 97, 99, 103, 108, 109, 110, 184, 188
Louis XV 147

M

Mahatma Gandhi 155, 190, 191
Marcel 27, 173
Marx 16, 51, 171
Matsuo Basho 125
Mead 22, 172
Mencius 88, 94, 96, 181
Meng Haoran 133
Mill 104, 110, 111, 112, 113, 114, 115, 117, 184
Miyake Setsurei 120
Mozart 83, 180
Muhammad 166

N

Nasr 156, 190
Nietzsche 9, 41, 82, 141, 168, 191

P

Peirce 21
Petöfi 136
Plato 7, 14, 44, 76, 85, 102, 105, 116, 171, 182, 183
Poe 130, 136
Pontius Pilate 163
Putnam 20, 21, 171, 172

Q

Qin Shihuang 44

R

Rameau 105, 141, 142, 143, 144, 145, 146, 149, 150, 152, 153, 187, 189, 190
Rawls 10, 85, 91, 92, 94, 99, 181
Roche 167, 191
Rosenstock-Huessy 5, 6, 9, 10, 13, 35, 36, 37, 38, 39, 40, 41, 42, 43, 44, 45, 47, 48, 49, 50, 51, 174, 175, 197
Rosenzweig 9, 10, 40, 42, 48, 51, 174, 197
Rousseau 97, 103, 107, 109, 110, 111, 123, 124, 125, 184

S

Saigyo 128, 132
Saint Augustine 146, 189
Saito Ryoku 128
Schelling 40
Schiller 69, 122, 123
Schlegel 11, 123, 171
Schopenhauer 124, 129
Shakespeare 128, 137
Shelley 127, 131, 136, 186, 187
Shiga Shigetaka 120
Shimazaki Toson 127, 137, 187
Stravinsky 74
Syed Hossein 156

T

Taylor 168, 191
Tocqueville 9, 111, 113
Togawa Shukotsu 121, 122, 187
Tokutomi Soho 120
Tolstoy 127, 180
Turgenev 127

V

Volland 145, 150, 188, 189, 190
Voltaire 142

W

Wang Wei 133
Whitehead 19, 171
Whitman 114
William 20, 22, 171, 184
Wing-tsit Chan 71, 179
Wordsworth 129, 132, 136

X

Xunzi 94, 181

Y

Yamaji Aizan 126

Z

Zilu 89

About the Contributors

Robert Allinson is Professor of Philosophy and the former Director of Humanities at Soka University of America (SUA). His numerous books and articles include *Saving Human Lives, Chuang-Tzu for Spiritual Transformation*, which has been translated into Chinese and Korean; *Global Disasters, Contemporary Perspectives, East and West*, and *Understanding the Chinese Mind*.

Roger T. Ames is Professor of Philosophy at the University of Hawaii and editor of *Philosophy East & West*. He has translated several Chinese classics, including *Sun-tzu: The Art of Warfare*, the *Confucian Analects* and the *Classic of Family Reverence: A Philosophical Translation of the* Xiaojing (both with H. Rosemont), and *A Philosophical Translation of the Daodejing: Making This Life Significant* (with D.L. Hall). He is also the author of *Thinking Through Confucius, Anticipating China: Thinking Through the Narratives of Chinese and Western Culture*, and *Thinking From the Han: Self, Truth, and Transcendence in Chinese and Western Culture* (all with D.L. Hall), and *Democracy of the Dead: Dewey, Confucius, and the Hope for Democracy in China* (with D.L. Hall). He is presently compiling the new *Blackwell Sourcebook of Chinese Philosophy*.

Miran Bozovic is Professor of Philosophy at the University of Ljubljana, Slovenia. In addition to many articles and chapters he is the author of *Der grosse Andere: Gotteskonzepte in der Philosophie der Neuzeit, An Utterly Dark Spot: Gaze and Body in Early Modern Philosophy*, and editor of Jeremy Bentham, *The Panopticon Writings*.

Wayne Cristaudo is Professor of Politics at Charles Darwin University NT, Australia. He is the author and editor of numerous books, journals and articles including *Power, Love and Evil: Contribution to a Philosophy of the Damaged* (which has been translated into Chinese), *Religion, Redemption, and Revolution: The New Speech Thinking of Franz Rosenzweig and Eugen Rosenstock-Huessy*, and *A Philosophical*

History of Love, and editor (with Paul Caringella and Glenn Hughes) of *Revolutions: Finished and Unfinished from Primal to Final*.

Alexander Dolin is Professor of Comparative Culture, World Civilization and Japanese Literature at Akita International University. He is the author of more than a dozen comprehensive monographs on Japanese literature and culture, Russian literature, culture and society as well as on the world civilizations issued in Russian, German, English and Japanese. He has also translated many volumes of classic and modern Japanese literature (especially poetry). His books include *Essays on Modern Japanese Poetry* (現代詩論). [in Russian with English Summary], *Kempo: The Tradition of Martial Arts*, [in Russian] which has over 700 000 copies in circulation, *A Prophet in his on Country: Prophetic, Messianic and Eschatological trends in Russian poetry, philosophy and social thought, 18-early 20 cc.*, (in Russian with English and Japanese summary), and *The Silver Age of Japanese Poetry*.

Waddick Doyle is the Director, Division of Global Communications and Film, and founder and director of the Masters in Global Communications program at AUP. Doyle teaches courses in Media Globalization, Contemporary World Television, Media Law, Policy and Ethics. He has held positions at universities in Italy, France and Australia.

His work covers the deeper cultural effects linked to transformations of media systems, and the development of a globalized brand media culture. He has published on what he calls the sacralisation of brands and reality television, and on media and belief. He is presently writing a book about the rise to power of the Italian media and advertising tycoon, Silvio Berlusconi, former Prime Minister of Italy.

Hélène Landemore (Ph.D. Harvard University, 2008) is Assistant Professor of Political Science at Yale University. Hélène's previous work includes a book in French on the notions of probability and reasonable choice in Hume and articles published in the *Journal of Moral Philosophy, Critical Review, Raison Publique,* and *Political Psychology*. She is also the co-editor (with Jon Elster) of *Collective Wisdom*.

Dr. Han Rui obtained her Master in Politics at the University of Warwick (2004) and her PhD in Political Philosophy at the University of Hong Kong (2010). She was visiting assistant professor at the Philosophy Dept. of HKU, Sept. 2010 - Jan. 2011, and is now teaching at Guangdong University of Foreign Studies. Her main research interest is normative political philosophy, esp. theories of justice. Her published papers include "Justice and Equality: An Introduction to Western Social Justice Theories", *Open Times*, No. 8 2010; and "Freedom of Speech, Equality, and the American Campaign Finance", *Open Times*, No. 4 2004. She was vice editor of *Re-reading America: Changes and Challenges*, Reardon's Publishing, UK, April 2004. She was also the translator of Harry V. Jaffa's *Crisis of the House Divided - An Interpretation of the Issues in the Lincoln-Douglas Debates*.

Donald Sturgeon is a PhD candidate at the department of philosophy, Hong Kong University. His main interests are in the fields of early Chinese thought, philosophy of language and epistemology, and he is currently working on early Chinese conceptions of knowledge. His publications include *Zhuangzi, perspectives, and greater knowledge*, as well as various works published in Chinese under his Chinese name 德龍 including 〈古漢語與墨辯的語言論〉 (*Classical Chinese and the theory of language of the Mohist Dialectical Chapters*) and 《墨辯中的語言哲學》 (*Philosophy of Language in the Mohist Dialectical Chapters*). He also edits and maintains the *Chinese Text Project* (http://ctext.org), an online database of classical Chinese literature.

Sun Youzhong is Assistant President, and the Dean of the School of English and International Studies at Beijing Foreign Studies University. He is the author of *Decoding China's Image: A Comparative Study of the China Reporting by The New York Times and The Times 1993-2002*; *John Dewey's Social Thought*; and co-author of *Modern American Popular Culture*; *Approaching America*; and *American Cultural Industry*. He is the editor of *English Education and Liberal Arts Education*; *Classics of Western Thought*; *Intercultural Mass Communication: Approaches to Key Texts in Cultural Studies*; *Intercultural Perspectives*; *Intercultural Communication Studies: New Frontiers* and *Cultural Studies Reader Series*, and co-translator of *Individualism Old and New: Selected Works of John Dewey*.

Heung Wah, Wong (王向华), Director of Global Creative Industries at the School of Modern Languages and Cultures, The University of Hong Kong. He holds a D. Phil from Oxford University. His career started with an anthropological study of a Japanese supermarket in Hong Kong and since 1990s he has worked on Western social thought, Japanese companies and management, globalization of Japanese adult videos, gender and sexualities in Taiwan, Japanese companies in Greater China, creative industries in East Asia, and international relations from an anthropological perspective. He is the author of *Japanese Bosses, Chinese Workers: Power and Control in a Hong Kong Megastore*, the co-author of *Japanese Adult Videos in Taiwan*, the co-editor of *From Faith in Reason to Reason in Faith: Transformations in Philosophical Theology from the Eighteenth to the Twentieth Century*, *St Augustine: His Legacy and Relevance*, *Dismantling the East-West Dichotomy: Essays in Honour of Jan Van Bremen*. His new co-authored books *Popular Culture and The Formation Hong Kong Identity* and *The Japanese Adult Video Industry* are forthcoming. He has also published a number of articles and book chapters on issues concerned with Japanese adult videos in Taiwan, Japanese overseas companies and their management, the globalization of Japanese popular culture, and international relations.